FULLY TUNED RADIAL BASIS FUNCTION NEURAL NETWORKS FOR FLIGHT CONTROL

The Kluwer International Series on
ASIAN STUDIES IN COMPUTER AND INFORMATION SCIENCE

Series Editor
Kai-Yuan Cai
Beijing University of Aeronautics and Astronautics, Beijing, CHINA

Editorial Advisory Board
Han-Fu Chen, Institute of System Science, Chinese Academy of Sciences
Jun-Liang Chen, Beijing University of Post and Telecommunication
Lin Huang, Peking University
Wei Li, Beijing University of Aeronautics and Astronautics
Hui-Min Lin, Institute of Software Technology, Chinese Academy of Sciences
Zhi-Yong Liu, Institute of Computing Technology, Chinese Academy of Sciences
Ru-Qian Lu, Institute of Mathematics, Chinese Academy of Sciences
Shi-Tuan Shen, Beijing University of Aeronautics and Astronautics
Qing-Yun Shi, Peking University
You-Xian Sun, Zhejiang University
Lian-Hua Xiao, National Natural Science Foundation of China
Xiao-Hu You, Southeast University
Bo Zhang, Tsinghua University
Da-Zhong Zheng, Tsinghua University
Bing-Kun Zhou, Tsinghua University
Xing-Ming Zhou, Changsha University of Technology

Also in the Series:

NONLINEAR CONTROL SYSTEMS AND POWER SYSTEM DYNAMICS by *Qiang Lu, Yuanzhang Sun, Shengwei Mei*; ISBN: 0-7923-7312-X

DATA MANAGEMENT AND INTERNET COMPUTING FOR IMAGE/PATTERN ANALYSIS *David Zhang, Xiobo Li and Zhiyong Liu* ISBN: 0-7923-7456-8

COMMON WAVEFORM ANALYSIS: *A New and Practical Generalization of Fourier Analysis*, by *Yuchuan Wei and Qishan Zhang;* ISBN: 0-7923-7905-5

DOMAIN MODELING-BASED SOFTWARE ENGINEERING: *A Formal Approach*, by *Ruqian Lu and Zhi Jin;* ISBN: 0-7923-7889-X

AUTOMATED BIOMETRICS: *Technologies and Systems*, by *David D. Zhang*; ISBN: 0-7923-7856-3

FUZZY LOGIC AND SOFT COMPUTING, by *Guoqing Chen, Mingsheng Ying Kai-Yuan Cai*; ISBN: 0-7923-8650-7

INTELLIGENT BUILDING SYSTEMS, by *Albert Ting-pat So, Wai Lok Chan*; ISBN: 0-7923-8491-1

PERFORMANCE EVALUATION, PREDICTION AND VISUALIZATION OF PARALLEL SYSTEMS by *Xingfu Wu*; ISBN: 0-7923-8462-8

ROBUST MODEL-BASED FAULT DIAGNOSIS FOR DYNAMIC SYSTEMS by *Jie Chen and Ron J. Patton*; ISBN: 0-7923-8411-3

FUZZY LOGIC IN DATA MODELING: *Semantics, Constraints, and Database Design* by *Guoqing Chen;* ISBN: 0-7923-8253-6

FULLY TUNED RADIAL BASIS FUNCTION NEURAL NETWORKS FOR FLIGHT CONTROL

by

N. Sundararajan

P. Saratchandran

Yan Li

*Nanyang Technological University
Singapore*

KLUWER ACADEMIC PUBLISHERS
Boston / Dordrecht / London

Distributors for North, Central and South America:
Kluwer Academic Publishers
101 Philip Drive
Assinippi Park
Norwell, Massachusetts 02061 USA
Telephone (781) 871-6600
Fax (781) 681-9045
E-Mail <kluwer@wkap.com>

Distributors for all other countries:
Kluwer Academic Publishers Group
Distribution Centre
Post Office Box 322
3300 AH Dordrecht, THE NETHERLANDS
Telephone 31 78 6392 392
Fax 31 78 6392 254
E-Mail <services@wkap.nl>

 Electronic Services <http://www.wkap.nl>

Library of Congress Cataloging-in-Publication Data

Sundararajan, N.
  Fully tuned radial bases function neural networks for flight control / N. Sundararajan, P. Saratchandran, Yan Li.
    p. cm. – (The Kluwer international series on Asian studies in computer and information science ; 12)
  Includes bibliographical references and index.
  ISBN 0-7923-7518-1 (alk. Paper)
  1. Flight control—Data processing. 2. Airplanes—Automatic control. 3. Neural networks (Computer science) 4. Adaptive control systems. I. Saratchandran, P. II. Li, Yan, 1972- III. Title IV. Series.

TL589.4 .S85 2001
629.132'6–dc21

2001038477

Copyright © 2002 by Kluwer Academic Publishers

All rights reserved. No part of this publication may be reproduced, stored in a retrieval system or transmitted in any form or by any means, mechanical, photo-copying, recording, or otherwise, without the prior written permission of the publisher, Kluwer Academic Publishers, 101 Philip Drive, Assinippi Park, Norwell, Massachusetts 02061

Printed on acid-free paper. Printed in the United States of America

To
My grandparents/parents
My wife Saraswathi
and my Daughters Sowmya and Priya

                        N.Sundararajan

To
My parents
My wife Jaya
and my children Para and Hemanth

                        P. Saratchandran

To
My parents
my husband Zhong XUE
and my baby LeQing XUE

                        Yan LI

SERIES EDITOR'S ACKNOWLEDGMENTS

I am pleased to acknowledge the assistance to the editorial work by Beijing University of Aeronautics and Astronautics and the National Natural Science Foundation of China

Kai-Yuan Cai
Series Editor
Department of Automatic Control
Beijing University of Aeronautics and Astronautics
Beijing 100083
China

Contents

Preface		xi
1. A REVIEW OF NONLINEAR ADAPTIVE NEURAL CONTROL SCHEMES		1
1.1	Adaptive Control of Nonlinear Systems Using Artificial Neural Networks	1
	1.1.1 Introduction	1
	1.1.2 An Overview of Neural Networks in Adaptive Control	4
	1.1.3 Nonlinear System Identification and Indirect Adaptive Control	8
	1.1.4 Direct Adaptive Control Strategy	9
1.2	A Review of Neuro-Flight-Control Strategies	12
	1.2.1 Autopilot Neural Flight Control Systems	14
	1.2.2 Fault Tolerant Neural Flight Control System	15
	1.2.3 High Angle of Attack Supermaneuver	17
1.3	Radial Basis Function Neural Network	18
1.4	Aircraft Flight Control Applications Using RBF Network	21

Part I Nonlinear System Identification and Indirect Adaptive Control Schemes

2. NONLINEAR SYSTEM IDENTIFICATION USING LYAPUNOV-BASED FULLY TUNED RBFN		29
2.1	Introduction	29
2.2	Stable Identification Using Lyapunov-Based Fully Tuned RBF Network	30
	2.2.1 Identification Strategy and System Error Dynamics	30
	2.2.2 Stable Parameter Tuning Rules	32
	2.2.3 Implementation of the Tuning Rule	34
	2.2.4 GRBF Network and Dead Zone Design	36
2.3	Simulation Results	38

	2.3.1	Example 1: Identification of SISO Nonlinear System	38
	2.3.2	Example 2: Identification of Nonlinear Time-Varying Missile System	40
2.4	Summary		44

3. REAL-TIME IDENTIFICATION OF NONLINEAR SYSTEMS USING MRAN/EMRAN ALGORITHM — 47
   3.1 Introduction — 47
   3.2 Introduction of MRAN Algorithm — 48
   3.3 Performance of the MRAN Algorithm — 52
       3.3.1 The ONSAHL Algorithm — 52
       3.3.2 BM-1: Nonlinear SISO Time-Invariant System — 54
       3.3.3 BM-2: Nonlinear SISO Time-Varying System — 56
   3.4 Real-Time Implementation of MRAN and the Extended MRAN Algorithm — 58
       3.4.1 Analysis of MRAN for Real-Time Implementation — 58
       3.4.2 BM-3: Nonlinear MIMO Time-Invariant System — 59
       3.4.3 Extended MRAN (EMRAN) Algorithm — 61
   3.5 Performance Comparison of MRAN *vs.* EMRAN — 64
       3.5.1 BM-2: Nonlinear SISO Time-Varying System — 65
       3.5.2 BM-3: MIMO Nonlinear Dynamic System — 66
   3.6 Summary — 67

4. INDIRECT ADAPTIVE CONTROL USING FULLY TUNED RBFN — 69
   4.1 Neural Network Based Indirect Adaptive Control — 69
   4.2 Neural Network Controller with Off-Line Training and On-Line Control — 71
       4.2.1 Linearized Longitudinal F8 Aircraft Model — 71
       4.2.2 Evolution of Off-Line Training — 72
       4.2.3 Off-Line Training/On-Line Control for the Linearized F8 Aircraft Model — 76
   4.3 On-Line Learning/On-Line Control for the Linearized Aircraft Model — 77
   4.4 Summary — 79

Part II Direct Adaptive Control Strategy and Fighter aircraft applications

5. DIRECT ADAPTIVE NEURO FLIGHT CONTROLLER USING FULLY TUNED RBFN — 85
   5.1 Overview — 85
   5.2 Problem Formulation — 86
   5.3 Stable Tuning Rule Using Fully Tuned RBFN — 87
       5.3.1 Control Strategy — 87

Contents ix

	5.3.2	RBFN Approximation and Error Dynamics	89
	5.3.3	Stable Adaptive Tuning Rule for Fully Tuned RBFN	90
5.4	Robustness Analysis		91
5.5	Implementation of the Tuning Rule		93
5.6	Summary		94

6. AIRCRAFT FLIGHT CONTROL APPLICATIONS USING DIRECT ADAPTIVE NFC ... 95
   - 6.1 Robust NFC for a Linearized F8 Aircraft Model ... 95
   - 6.2 NFC for Localized Nonlinear Aircraft Model ... 107
     - 6.2.1 Localized Nonlinear Fighter Aircraft Model ... 107
     - 6.2.2 Performance Evaluation of the NFC ... 108
     - 6.2.3 Discussion ... 111
   - 6.3 NFC for Full-Fledged Nonlinear 6-DOF Aircraft Model ... 111
     - 6.3.1 Full-Fledged Nonlinear 6-DOF Aircraft Model ... 112
     - 6.3.2 Kinematic and Navigation Equations ... 113
     - 6.3.3 Auxiliary Equations ... 115
     - 6.3.4 Other Equations ... 116
     - 6.3.5 High α Stability-Axis Roll Maneuver ... 119
     - 6.3.6 Simulation Studies ... 121
   - 6.4 Summary ... 125

7. MRAN NEURO-FLIGHT-CONTROLLER FOR ROBUST AIRCRAFT CONTROL ... 127
   - 7.1 Problem Formulation and Conventional Controller ... 127
   - 7.2 Robust MRAN-Flight-Controller ... 128
   - 7.3 Simulation Results ... 130
     - 7.3.1 Response to Model Error ... 131
     - 7.3.2 Response to Actuator Sluggishness ... 136
   - 7.4 Summary ... 139

8. CONCLUSIONS AND FUTURE WORK ... 141
   - 8.1 Conclusions ... 141
   - 8.2 Recommendations for Further Research ... 143

Bibliography ... 145

Topic Index ... 157

Preface

Purpose and Goals

In the past three decades, major advances have been made in adaptive control of linear time-invariant plants with unknown parameters. The choice of the controller structure is based on well established results in linear systems theory, and stable adaptive laws which assure the global stability of the overall systems are derived based on the properties of those systems. In contrast to this, mature design procedures that simultaneously meet the requirements of stability, robustness, and good dynamic response for nonlinear system control are currently not available.

Recently, Artificial Neural Network (ANN) based control strategies have attracted much attention because of their powerful ability to approximate continuous nonlinear functions. Specifically, a neural controller with on-line learning can adapt to the changes in system dynamics and hence is an ideal choice for controlling highly nonlinear systems with uncertainty. Among a variety of network structures, Radial Basis Function Network (RBFN) has been studied intensively due to its good generalization ability and a simple network structure that avoids unnecessary and lengthy calculations . All the advantages of the RBFN have motivated us to further investigate its use in the area of nonlinear adaptive control in this book, with emphasis in aircraft flight control applications.

The classical approach for Gaussian RBFN implementation is to fix the number of hidden neurons, centers and widths of the Gaussian function *a priori*, and then estimate the weights connecting the hidden and output layers using parameter tuning rules, like LMS, RLS etc. However, in practice it is difficult to choose the centers and widths appropriately, especially for on-line implementation where preliminary training is impossible. The inaccurate centers and widths will unavoidably result in the deterioration of the performance, especially when coping with highly nonlinear systems with uncertainty, such as robot, aircraft, etc. In comparison to conventional approaches, recently fully tuned RBFNs have shown their potential for accurate identification and control. In a fully tuned RBFN, not only the weights of the output layer, but also the other parameters of the network (like the centers and widths) are updated, so that the nonlinearities of the dynamic system can be captured as quickly as possible.

In this book, we first address the theoretical aspects of designing stable nonlinear adaptive control law with a fully tuned RBFN, and then explore the applications of the controllers designed for aircraft flight control. More specifically, the objectives of the book can be summarized as:

- To design indirect adaptive control and direct adaptive control strategies incorporating fully tuned RBFN networks. In the indirect control strategy, a stable identification scheme using the fully tuned RBFN is developed for identification of nonlinear systems with external inputs. In the direct adaptive control scheme, the objective is to design the on-line control law based on a fully tuned RBFN, guaranteeing the stability of the overall system.

- To explore the applications of the proposed neuro-controller in the field of aircraft flight control. Simulation studies are carried out based on different control objectives and aircraft models, including command following for a linearized F8 aircraft model in longitudinal mode, pitch-rate control for a localized nonlinear fighter aircraft model, and implementing a high α stability-axis roll maneuver based on a full-fledged 6-DOF high performance aircraft model with nonlinear dynamic nature.

- To evaluate the recently developed MRAN algorithm for real-time nonlinear system identification and adaptive control, especially in fault tolerant aircraft flight control applications.

An Overview

An overview of the main contributions made in this book are:

- A new stable identification scheme based on a fully tuned Growing RBFN (GRBFN) is developed for identification of nonlinear systems with external inputs, which extends the existing schemes of only tuning the weights of the RBFN. The proposed method not only guarantees the stability of the

PREFACE

overall system, but also improves the performance for the identification. This identification scheme is then used in an indirect adaptive control setting.

- A new direct adaptive control scheme using the fully tuned RBFN is developed for nonlinear system control. This approach extends Kawato's conventional feedback-error-learning where only the weights of the RBFN controller are adaptable. The tuning rule for updating all the parameters of the RBFN is derived based on the Lyapunov stability theory, guaranteeing the stability of the overall system. The robustness of the proposed neuro-controller is analyzed in terms of approximation errors and model errors. By tuning all the parameters of the network on-line, there is no need to estimate the centers and widths of the Gaussian functions embedded in the RBFN controller *a priori*, resulting in a better tracking performance.

- In this book, the applications of the proposed neuro-controller scheme to aircraft flight control are studied in detail. To accomplish this, simulation studies are carried out based on different control objectives and aircraft models, including command following for a linearized F8 aircraft in longitudinal model, pitch-rate control for a localized nonlinear fighter aircraft model, and most prominently, implementing a high α stability-axis roll maneuver based on a full-fledged 6-DOF nonlinear high performance fighter aircraft model. The simulation results demonstrate the superior performance of the proposed neuro-controller scheme, validating the theoretical results derived earlier.

- Several sequential learning algorithms for implementing a fully tuned RBF network are investigated, including the MRAN algorithm. Based on a runtime analysis of the MRAN algorithm for real-time identification, a new algorithm called extended MRAN (EMRAN) is proposed in this book. Simulation studies based on benchmark problems demonstrate that by incorporating a "winner neuron" strategy to the existing MRAN algorithm, the EMRAN algorithm can improve the learning speed greatly with the accuracy close to that of the MRAN.

- MRAN is used for the first time as a fault-tolerant controller for controlling a linearized F8 aircraft model. Although this approach lacks a strict mathematical proof, it is demonstrated from the simulation results that the MRAN controller can implement a more compact network structure with improved tracking accuracy.

Organization of the Book

The book is organized as follows.

Chapter 1 presents a detailed review of the RBFN in the field of nonlinear adaptive control, with an emphasis in aircraft flight control applications.

The rest of this book has been divided into two parts. Part I investigates the indirect adaptive control scheme using the fully tuned RBFN and consists of three chapters. In Chapter 2, different sequential learning algorithms are evaluated for nonlinear system identification. A new stable identification scheme is developed, and a stable tuning law is derived using Lyapunov method. Chapter 3 presents other learning algorithms, including the Minimal Resource Allocation Network (MRAN) for the identification of nonlinear systems. A new algorithm called Extended MRAN (EMRAN) to increase the on-line learning speed is also developed in this chapter. Using the proposed identification schemes, the performance of the indirect adaptive control is evaluated in Chapter 4.

Part II focuses on the development of the direct adaptive control strategies using the RBFN and their applications to aircraft flight control, and it contains three chapters. A new on-line neuro-control scheme including the parameter adjusting rule for the fully tuned RBFN controller is derived in Chapter 5. In Chapter 6, simulation studies demonstrating the effectiveness of the proposed method is presented based on several aircraft fighter models, varying from the linearized longitudinal F8 aircraft model, to a full-fledged nonlinear 6-DOF high performance fighter aircraft model. Chapter 7 presents the use of MRAN as a fault tolerant controller in aircraft control application for the first time. Chapter 8 provides a summary of the book with possible future directions.

Acknowledgements

We wish to acknowledge the encouragement and support of many individuals who made this task possible.

First and foremost, we wish to thank Dr. Cham Tao Soon, President, Nanyang Technological University(NTU), Singapore for providing an excellent academic and research environment which is a necessary prerequisite for an endeavor such as this to succeed.

We are grateful to Prof. Er Meng Hwa, Dean, School of Electrical and Electronic Engineering, and Prof. Soh Yeng Chai, Head, Control and Instrumentation Division in NTU for their support during this work.

Special thanks are due to Mr. Zhong XUE who helped us in bringing the book in the final form.

PREFACE

We owe a debt of gratitude to Prof. Kai-Yuan CAI, Department of Automatic Control, Beijing University of Aeronautics and Astronautics,Beijing, 100083, China, who encouraged us to write this book. Thanks are also due to the anonymous reviewers for their valuable comments, most of which have been incorporated in the book.

Finally, we extend our thanks to Ms. Melissa Fearon, Kluwer Academic Publishers, Norwell, MA, USA for extending her full cooperation and support in this effort.

Singapore   P.S
May 2001.   N.S
                                                        Y.L

Chapter 1

A REVIEW OF NONLINEAR ADAPTIVE NEURAL CONTROL SCHEMES

Over the last four decades, adaptive control has evolved as a powerful methodology for designing feedback controllers of nonlinear systems. However, most of these studies assume that the system nonlinearities are known *a priori*, which is generally not applicable in the real world. To overcome this drawback, from 1990s, there has been a tremendous amount of activity in applying Neural Networks (NNs) for adaptive control. With their powerful ability to approximate nonlinear functions, neuro-controllers can implement the expected objectives by canceling or learning the unknown nonlinearities of the systems to be controlled. NNs are especially suitable for the adaptive flight control applications where the system dynamics are dominated by the unknown nonlinearities. Moreover, among different choices of network structures, Radial Basis Function Network (RBFN) has shown its potential for on-line identification and control, and hence arouses much research interest.

In this chapter, a brief introduction on adaptive control is given, followed by a detailed review of adaptive neural control strategies, emphasizing the aircraft flight control applications. Then, we concentrate in exploring the applications of the RBFN in adaptive control and highlight its status in aircraft flight control. The last section discusses the problems existing in the current research work.

1. Adaptive Control of Nonlinear Systems Using Artificial Neural Networks

1.1. Introduction

Adaptive control was motivated by the problem of designing autopilots for aircraft operating at a wide range of speeds and altitudes in 1950s. However, it was only in the 1960s, when Lyapunov's stability theory was firmly established, the convergence of the proposed adaptive control schemes can be mathematically proven. Following this, in 1970s, complete proofs of stability for several adaptive schemes appeared. Further, in the late 1970s and early 1980s, state space based proofs of stability for model reference adaptive schemes appeared in the works of Narendra, Lin, Valavani [82] and Morse [69]. Surveys of the applications of adaptive control are given in books edited by Narendra and Monopoli [83] and Unbehauen [120]. Those early results on stability dealt with the ideal case of no modeling errors — the only uncertainty in the system being due to unknown parameters. Since then, a considerable amount of adaptive control research has been devoted to the development of robust adaptive control systems, where closed-loop stability properties are retained in the presence of not only a large parametric uncertainty, but also of modeling errors [40][39]. On the other hand, in the 1970s and 1980s, there also has been a great deal of interest in the use of state feedback to exactly linearize the input-output behavior of nonlinear control systems. The theory of linearization by exact feedback was developed through the efforts of several researchers, such as Singh an Rugh [112], Isidori, Krener, Gori-Giorgi and Monaco [41]. Good surveys are available in [24], and a number of applications of these techniques are illustrated in [109]. All these efforts in research and development has given substantial innovations and the fundamental issues of adaptive control for linear systems, such as selection of controller structures, development of adaptive law, etc., have been well established. These results have been reported in several books dealing with the design and analysis of adaptive systems, for example, [80][40][108][4].

Although it is known that most practical systems are inherently nonlinear, adaptive control of such systems was not seriously considered until the advances in geometric nonlinear systems theory, such as feedback linearization, were developed. With these achievements, in the last ten years, adaptive control has evolved as a rigorous control strategy for a reasonably large class of nonlinear systems. Many remarkable results and new design tools, such as backstepping procedure, feedback linearization techniques, have appeared in both Lyapunov and estimation based schemes [45][47]. Some of these methods are briefly reviewed.

- **Dynamic Inversion**

    Dynamic inversion is a technique for control law design in which feedback is used to simultaneously cancel system dynamics and achieve desired response characteristics. Consider a first order affine system:

    $$\dot{\mathbf{x}} = \mathbf{f}(\mathbf{x}) + \mathbf{G}(\mathbf{x})\mathbf{u} \tag{1.1}$$

    $\mathbf{x}$ is the state vector, the control law $\mathbf{u}$ which yields the desired $\dot{\mathbf{x}}_d$ is,

    $$\mathbf{u} = \mathbf{G}^{-1}(\mathbf{x})(-\mathbf{f}(\mathbf{x}) + \dot{\mathbf{x}}_d) \tag{1.2}$$

In this way, dynamic inversion control laws present an attractive alternative to conventional gain-scheduled designs for systems with complicated nonlinearities. However, the desired dynamics is obtained if the inversion is exact and this can occur only with an exact knowledge of $\mathbf{f}(\mathbf{x})$ and $\mathbf{G}(\mathbf{x})$. In practice, assuming exact knowledge of the system dynamics can never be the case and robustness to modeling errors is a serious problem. Some applications to aircraft flight control using nonlinear inversions are [6],[13],[17],[115].

- **Feedback Linearization**
Feedback linearization is an approach to nonlinear control design which has attracted a great deal of research interest in recent years. The central idea of the approach is to algebraically transform a nonlinear system dynamics into a fully or partially linear one, so that linear control techniques can be applied [114].

To describe the basic concepts of feedback linearization, consider a simple single-input single-output (SISO) nonlinear systems with a controllable canonical form,

$$x^{(n)} = f(\mathbf{x}) + g(\mathbf{x})u \quad (1.3)$$

where u is the scalar control input, $x^{(n)}$ is the nth order derivative of x, which represents the scalar output of interest. $\mathbf{x} = [x, \dot{x}, \cdots, x^{(n-1)}]^T$ is the state vector, and $f(\mathbf{x})$ refers to a nonlinear function of the states. Eq.(1.3) can be written using state-space representation,

$$\frac{d}{dt}\begin{bmatrix} \dot{x}_1 \\ \vdots \\ \dot{x}_{n-1} \\ \dot{x}_n \end{bmatrix} = \begin{bmatrix} x_2 \\ \vdots \\ x_n \\ f(\mathbf{x}) + g(\mathbf{x})u \end{bmatrix} \quad (1.4)$$

For systems which can be expressed in the controllable canonical form, let the control input (assuming g to be non-zero) as,

$$u = \frac{1}{g}(v - f) \quad (1.5)$$

we can cancel the nonlinearities and obtain an input-output relation (multiple integrator form): $x^{(n)} = v$. For tasks involving the tracking of a desired output $x_d(t)$, the control law,

$$v = x_d^{(n)} - k_1 e - k_2 \dot{e} - \cdots - k_{n-1} e^{(n-1)} \quad (1.6)$$

leads to an exponentially convergent tracking provided the k_i's are chosen in such a way that $p^n + k_{n-1}p^{(n-1)} + \cdots + k_1$ is a stable polynomial

($e = x(t) - x_d(t)$ is the tracking error). The feedback linearization can be extended to a multi-input multi-output (MIMO) system in a straightforward way, and papers that apply the feedback linearization in the control strategy can be found in [113][90].

Feedback linearization has been used successfully to address some practical control problems, including control of high performance aircraft, industrial robot. However, a key assumption in this study is also that the system nonlinearities are known *a priori*.

With the complexity of modern day systems increasing and the performance requirements for industrial and military systems becoming more stringent, it may be impossible to estimate the system nonlinearities using the traditional approach. In many applications, the conventional adaptive controllers alone are no longer sufficient to provide accurate enough control. Under these circumstances, new design tools and approaches have to be explored.

1.2. An Overview of Neural Networks in Adaptive Control

Artificial Neural Networks (ANNs) have emerged over the last three decades as one of the most vigorous and exciting fields of modern science, because of their powerful ability to learn, to generalize, and to adapt to unknown nonlinearities. In the last decade, almost independent from the adaptive nonlinear control research, a tremendous amount of activity in neurocontrol approach has been carried out [85][66][84].

Generally, for adaptive control purposes neural networks are used as approximation models of unknown nonlinearities. The input/output response of neural network models is modified by adjusting the values of its parameters. Although it is true that polynomials, trigonometric series, splines, and orthogonal functions can also be used as function approximator, neural networks have been found to be particularly useful for controlling highly uncertain, nonlinear, and complex systems [126][79][50].

Neural control strategies can be broadly classified into off-line and on-line schemes based on how the parameters of the network are tuned. When the neural controller operates in an on-line mode, it has no *a priori* knowledge of the system to be controlled, and the parameters of the network are updated while the input-output data is received. However, in the off-line control, the network's parameters are determined from the known training pairs, and then those parameters are fixed for control purposes.

Though the role of neural networks in the on-line and/or off-line learning are to learn some underlying functions related to the system, the methodologies differ by the way in which they are utilized. They range from learning the inverse dynamics of the system without any guarantee of stability [66], the feedback linearization strategy [15] to the direct adaptive control that guarantees stability

A Review of Nonlinear Adaptive Neural Control Schemes

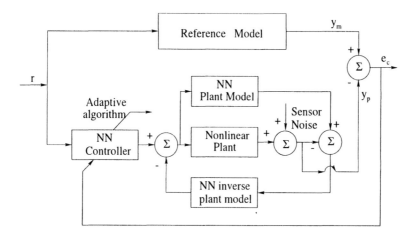

Figure 1.1. Adaptive inverse control system

of the overall systems [107][27]. In [1], Agarwal has given a comprehensive classification of the neural-network-based control schemes proposed in the literature. In Agarwal's work, the NN based control schemes are classified into two main categories; One is called "neural network only as an aid", where the neural networks are used to relieve the complex diagnostics, tedious modeling, unwieldy tuning, or excessive computational effort in conventional control schemes. The other class comprises schemes where a neural network is used as a controller and is updated using available input signals. The data pairs for adjusting the network are obtained either from the real plant, or using a model simulation.

It can be seen from the literature that neural network based adaptive control covers a very broad area, and in this section we introduce some of the popular control strategies. For details, one can refer to [1][30][46][37][51].

- **Adaptive Inverse Control [125]**

  Fig.1.1 shows a structure for the model reference adaptive inverse control proposed in [125]. The adaptive algorithm receives the tracking error e_c between the plant output y_p and the reference model output y_m. r is the desired output (or command input). The controller parameters are updated to minimize e_c. The basic model reference adaptive control approach may be affected by sensor noise and plant disturbances. An alternative which allows cancellation of the noise and disturbances includes a neural network model in parallel with the plant. That model will be trained to receive the same inputs as the plant and to produce the same output. The difference between the two outputs will be interpreted as the effect of the noise and

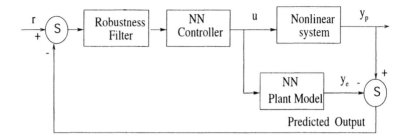

Figure 1.2. Nonlinear internal model control

disturbances at the plant output. It will then enter the inverse plant model to generate a filtered noise and disturbance signal that is subtracted from the plant inputs.

- **Nonlinear Internal Model Control [73]**
  Nonlinear internal model control, shown in Fig.1.2, consists of a neural network controller, a neural network plant model, and a robustness filter with a single tuning parameter. The neural network controller is generally trained to represent the inverse of the plant. The error between the output of the neural network plant model y_p and the measurement of plant output y_e is used as the feedback input to the robustness filter, which then feeds into the neural network controller. The NN plant model and the NN controller can be trained off-line, using data collected from plant operations. The robustness filter is a first order filter whose time constant is selected to ensure closed loop stability.

- **Neural Adaptive Feedback Linearization [52]**
  The neural adaptive feedback linearization technique is based on the standard feedback linearization controller introduced in the earlier section. An implementation is shown in Fig.1.3. The feedback linearization technique produces a control signal with two components. The first component cancels out the nonlinearities in the plant, and the second part is a linear state feedback controller. The class of nonlinear systems to which this technique can be applied is described by a class of affine dynamics as in Eq.(1.3). If we approximate the function f and g using the two neural networks, respectively, the control strategy of feedback linearization can be implemented. It should be noted that there are several variations on the neural adaptive feedback linearization controller, including the approximate models of Narendra [81]. For details, refer to [52][81].

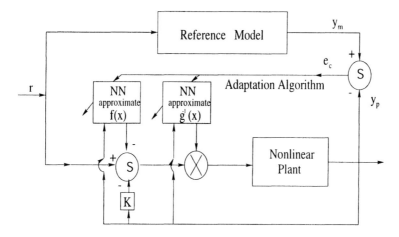

Figure 1.3. Neural adaptive feedback linearization

- **Stable Direct Adaptive Control [107]**
  Recently several direct adaptive control strategies have been designed to guarantee the overall system stability. For example, the method of [107] uses Lyapunov stability theory for designing the network tuning rule, rather than using a gradient descent algorithm like backpropagation. This control strategy has attracted much research interest because the system's stability can be guaranteed according to a strict mathematical proof.

Other NN based control schemes, such as model predictive control [37], adaptive critic [30], are also being studied. It would be impossible to describe in detail all the control strategies, instead we will concentrate on the most commonly used control strategies, namely, indirect adaptive control and direct adaptive control schemes. The adaptive inverse control, stable direct adaptive control, and fixed stabilizing controller [30] (i.e., feedback-error-learning controller) all belong to these two categories.

In indirect adaptive control, generally two NNs are used: one is for identifying the forward/inverse dynamics of the system, and the other is connected in cascade with the system to be controlled and its parameters are updated on-line to implement a suitable control law. In indirect schemes, the system identification procedure needs to be chosen with care [123]. While in direct adaptive control, there is no explicit attempt to determine a model of the process dynamics, and the parameters of the network controller are directly adjusted to reduce some norm of the output error.

8   FULLY TUNED RBF NEURAL NETWORK FOR FLIGHT CONTROL

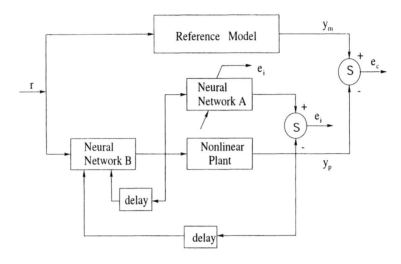

Figure 1.4.   Indirect adaptive control scheme (learns the system dynamics)

On the other hand, in applying the neural networks in adaptive control, the Multilayer Feedforward Network (MFN), Multilayer Perceptron (MLP) [85][105] and Radial Basis Function Network (RBFN) [107][31][95] have become the most popular network architectures. The varied network structures coupled with different control schemes, different learning algorithms for the network constitutes the big family of NN based control strategies. In the following sections, both the indirect and direct adaptive control strategies are surveyed, with the objective to find out a proper on-line scheme and effective parameter tuning rule for nonlinear system control using RBFN, especially for aircraft flight control applications.

1.3. Nonlinear System Identification and Indirect Adaptive Control

Before Park *et al.* rigorously proved that RBFNs having one hidden layer are capable of universal approximation [92], MFN and MLP were used as the main network structures for the nonlinear system identification and the identification based indirect adaptive control schemes. In 1990, Narendra and Parthasarathy proposed several models for identifying and controlling of nonlinear dynamic systems [85]. Fig.1.4 shows NN based indirect control scheme. The objective of the control is to track the desired output y_m of the reference model. For this purpose the network A is learning the nonlinear plant's dynamics. An iterative off-line technique is employed to adjust the parameters of this network so as to develop the inverse system model of the plant (network B). To implement the network structure, a gradient method is used for updating the parameters, which is time-consuming and may lead to instability for large values of the step size. Since for real-time control, the neural network should work in an on-line mode to adapt its parameters to plant dynamics which is poorly understood or rapidly changing, it is essential that efforts be directed towards selecting better network structures and determining effective adaptive laws.

Around 1991, Billing and his co-workers, in a series of papers [21][22][19] have explored RBFNs for nonlinear dynamic system identification. Experimental results reveal that RBFN is more capable of nonlinear system identification than the traditional MFN. However, they did not extend the identification scheme to control applications. Soon after, a turning point in the use of RBFNs was brought out by Park *et al.* in 1991 [92], who rigorously proved the RBFN's universal approximation ability. That is, given enough hidden neurons, RBF network with one hidden layer can approximate any continuous functions with arbitrary accuracy. After this, there has been considerable interest in applying RBFN as the basic structure of NNs for nonlinear system identification and control. Behera *et al.* investigated the application of inversion of a RBF network to nonlinear control problems for which the structure of the nonlinearity is unknown [7]. Initially, the RBFN is trained to learn the forward dynamics of the plant, and then the control structure is proposed based on this identified RBFN model. A feedback control law is derived according to the input prediction by inversion of the RBFN model so that the system is Lyapunov stable and an EKF based algorithm is employed to carry out the network inversion during each sampling interval. Simulation studies on the single link robotic manipulator showed that the performance of the proposed method outperforms Narendra's method which is based on the gradient algorithm and needs the prior knowledge of the nonlinear system.

The structure of another kind of identification based indirect adaptive control strategy is shown in Fig.1.5. In this strategy, neural network A is used on-line to identify the inverse dynamics of the system directly, and neural network B, with the same parameters copied from network A, is connected in cascade with the system to be controlled as a "copy controller" to generate the desired control signal. In this way, there is no need to compute the inverse dynamics of the system according to the system's forward dynamics. The application of this inverse strategy for aircraft autopilot application can be found in [105].

1.4. Direct Adaptive Control Strategy

Much has been written on the system identification and identification based indirect adaptive control. Nevertheless, the indirect control strategy depends strongly on the identification procedure, and generally there is no strict mathematical proof to guarantee the stability of the tracking error and the convergence of the network parameters. In 1992, Sanner and Slotine proposed a direct

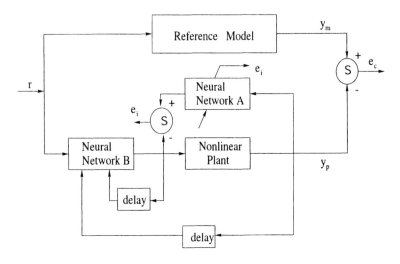

Figure 1.5. Indirect adaptive control scheme (learns the system's inverse dynamics)

adaptive tracking control scheme to solve these problems [107], the structure of which is shown in Fig.1.6. This architecture employees a Gaussian RBF network to adaptively compensate for the plant nonlinearities. Under mild assumptions about the degree of smoothness exhibited by the nonlinear functions, the algorithm is proven to be globally stable, with tracking errors converging to a neighborhood of zero. To implement the RBFN, a stable weight adjustment mechanism is derived using Lyapunov theory. With this tuning rule, the weights of the RBFN converge to the optimal weights gradually. The network construction and the performance of the resulting controller are illustrated through simulations with example systems. From Sanner's method we can see that to synthesize an adaptive controller with guaranteed stability, it is first necessary to ensure that the chosen architecture is capable, for some values of its adjustable parameters, of producing the control signals necessary to achieve the desired performance from the system being controlled.

Lewis *et al.* also proposed a neural network controller for a general serial-link rigid robot arm [52]. The neural network used in this strategy is the MFN and the rule for tuning the weights of the NN is derived based on a direct control strategy. Using Lyapunov stability theory, the derived tuning rule can guarantee the convergence of the tracking error and the stability of the overall system. In [26], the tracking control of robots in joint space using a RBFN controller is studied and an on-line direct control strategy is proposed to provide high tracking accuracy of robot path following performance. However, in these strategies, most of the works focus on the tracking performance while ignoring

A Review of Nonlinear Adaptive Neural Control Schemes

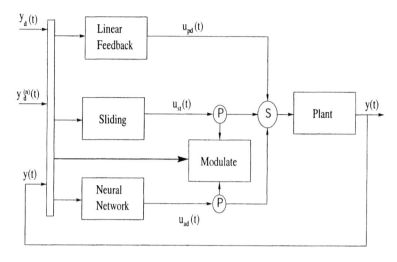

Figure 1.6. Direct adaptive control strategy (nonlinearity compensation)

the ability of the neural network to capture the underlying nonlinear function. To investigate the possibility of using NNs for on-line function approximation and control of a nonlinear dynamic system, Kawato proposed an on-line control strategy [27], where a Gaussian RBFN is used to learn the inverse dynamics of the system. The control strategy, with the name of "feedback-error-learning", is shown in Fig.1.7.

In the feedback-error-learning strategy, the total control effort, u, is composed of the output of the neural controller, u_{nn}, and the output of the conventional feedback controller, u_p. If the total control effort output by the RBFN does not reflect the desired control effort, the parameters of the neural controller must be adjusted according to some updating rules. In Kawato's strategy, the output of the conventional controller is utilized as the feedback error signal to tune the parameters of the RBFN, since it reflects the error in the control effort. In this formulation, it is expected that the output of the conventional feedback controller will tend to zero as the neural controller learns the appropriate control law.

Feedback-error-learning has the advantage of learning the true inverse dynamics without requiring that the network be trained initially off-line. By utilizing the on-line data, it allows the controller to track the system dynamics which may change along with time. This proposed control scheme has been utilized to control the movement of an industrial robotic manipulator, and it was demonstrated that the performance is good provided enough hidden neurons were chosen. It is worth noting that nowadays neural networks based direct

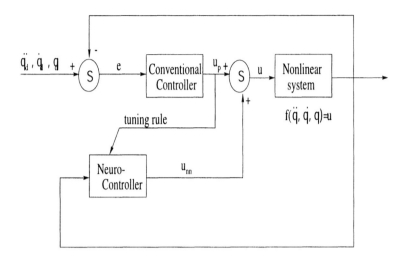

Figure 1.7. Feedback-error-learning control strategy

adaptive control strategies have attracted increasing attention, because the stability of the overall system can be guaranteed provided the adaptive tuning law for the nonlinear system is derived based on a Lyapunov synthesis approach.

So far, we have surveyed neuro-control schemes in adaptive control of nonlinear systems. In the next section, a detailed literature review emphasizing the use of neural networks for aircraft flight control applications is presented.

2. A Review of Neuro-Flight-Control Strategies

Aircraft flight control has been dominated by the classical control techniques. They have progressed from very simple fixed-form feedback structures with gains tuned by control engineers in flight, to complex multivariable feedback laws that are designed with modern multivariable tools. Over the last decade or so, control researchers have begun to apply flight control design methodology such as dynamic inversion [71] or feedback linearization [11], [38], [41] from prevailing paradigms common to many complicated engineering problems.

Although these conventional controllers have produced many highly reliable and effective control systems, next generation aircraft using Active Control Technology (ACT) may present control designers with a variety of unprecedented challenges. For example, in high angle-of-attack (AOA) flight, the aerodynamics are poorly understood and expensive to model. Alternatively, variation in dynamic response may occur due to battle damage or component failure, requiring rapid on-line reconfiguration of the control system to maintain stable flight and reasonable levels of handling qualities.

A Review of Nonlinear Adaptive Neural Control Schemes

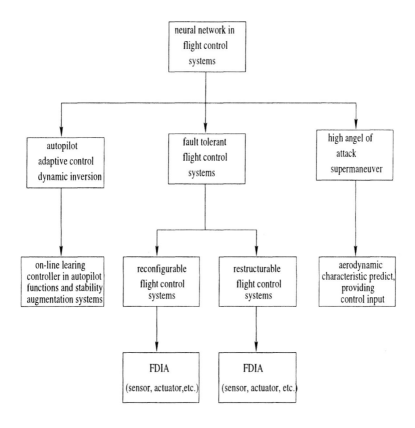

Figure 1.8. Applications of neural network in flight control

To accomplish these control objectives with gain scheduling, the nonlinear aircraft equations are linearized at several equilibrium flight conditions over the desired flight envelope. Then the control gain is designed for these different flight conditions. The optimal gains at each scheduling point should guarantee robust stability and performance; that is, they should guarantee stability and good performance at points near the designed equilibrium point. Since the feedback gains are scheduled *a priori*, no automatic corrective action is taken to mitigate the effects of a control law that is no longer appropriate. Hence, in case of time-varying dynamics and other unanticipated events, the performance deteriorates greatly. To give an example, when an aircraft has just released a significant quantity of external stores, there can be a very considerable change in its mass, such mass changes cause changes in the aircraft dynamics.

The primary difficulty associated with the use of feedback linearization or nonlinear inversion is also that a detailed knowledge of the nonlinear plant dynamics is required. A high fidelity nonlinear force and moment model must be constructed and inverted in real-time using an on-board computer, which is also a computationally intensive task. Moreover, in the future, the above control system design methods will become increasingly difficult with the growing complexity of new high performance aircraft. Therefore, it is expected that aircraft of the future will benefit from an adaptive control system based on the full nonlinear dynamics of the system while avoiding detailed knowledge of the system or prohibitively complex gain scheduling.

Due to their powerful ability of approximating nonlinear functions, control laws and design methods incorporating artificial neural networks have been extensively studied in the area of flight control. In [17], Calise *et al.* has summarized some current research efforts of applying NN technology for flight control system design, with emphasis in nonlinear adaptive control. It has been shown that NN with on-line learning can adapt to aircraft dynamics which is poorly known or rapidly changing. In addition, neural networks in the flight control architecture offer speed in performing the model inversion required to implement feedback linearization.

In the literature, there are a number of papers that have addressed flight control applications using NNs. Applications of neural networks to future aircraft technology was studied by Steiberg and DiGirolamo [116]. They outlined a number of potential areas in flight control for neural networks. In this review, NN applications for flight control systems are divided into three major areas as shown in Fig.1.8, namely, autopilot flight control system, fault tolerant flight control system, and high angle of attack supermaneuver system. However, it should be noted that in practice there is no distinct border among all these applications.

In the remainder of this section, a detailed review of the NN in aircraft flight control is presented based on the above classification.

2.1. Autopilot Neural Flight Control Systems

The implementation of on-line neurocontrollers in the autopilot control laws for high performance aircraft has the following advantages: first, it can avoid the precomputation, storing, and interpolation between thousands of feedback gains of a typical flight control system. Second, it has the ability to compensate for nonlinearities and model uncertainties. Last, through on-line learning ability, the designed controller avoids the time-consuming gain recalculation following any modification to the aircraft or its control system during the operative life.

Baker and Farrel [5] motivated and described a NN based design approach for the development of a learning augmented flight control system. When combined with adaptation, the resulting hybrid methodology yields a superior control strategy related to conventional techniques in terms of: design and tuning costs, accommodation of uncertainty, closed-loop system performance and operational efficiency. However, no flight control application was demonstrated. Troudet *et al.* [119] designed a model following neuro-controller for an integrated airframe/propulsion linearized model of a modern fighter aircraft to provide independent control of pitch rate and airspeed responses to pilot command inputs. Only the piloted longitudinal landing task was considered. Results are presented from a preliminary study of the neuro-control design, and the desirable tracking performance is obtained by the neuro-controller.

In [75], Napolitano and Kincheloe investigate the neural controllers in the control laws of the autopilot functions and of stability augmentation systems for both longitudinal and lateral directional dynamics. Extended Back-Propagation (EBP) was used as the training algorithm for the on-line learning neural controller, and its accuracy and learning speed were studied. The superiority of the EBP is demonstrated in various autopilot systems and in different flight conditions.

2.2. Fault Tolerant Neural Flight Control System

In the traditional flight control systems, a component failure is fixed for by one or several identical component backups. For example, F18 aircraft (USA), Mirage 400 (France), F-104G (USA) are all quadruple redundant. However, these redundancy levels imply a substantial weight penalty, and the increase of the components make the problem of safety even worse. Due to these reasons, increasing interest in developing robust and adaptive controller for aircraft without component redundancies has opened up the field of fault tolerant neural controller design, so as to regain the control of the aircraft under actuator lock, control surface loss, sensor failure, or a combination of them.

Fault tolerant flight control means that when an aircraft undergoing the failures, the controller's output can lead the partially damaged aircraft back to the equilibrium state, and even implement the scheduled target with a sacrifice in performance. In some papers, it is also referred to as Failure Detection, Identification and Accommodation (FDIA). Reconfiguration and restructuring are two ways to implement the FDIA, and there is a fine distinction between the two approaches. In reconfigurable flight control systems, the types of control failures are classified and anticipated, then the relative change in control gains are calculated off-line, using any selected control scheme, and stored in memory of flight control computer, ready for on-line use. On the contrary, in restructurable flight control systems, the reconstruction of the control law is performed on-line. In this case, the availability of sufficient computational power, as opposed to memory, can potentially become a critical issue.

In the field of FDIA using reconfigurable strategies, Napolitano and his colleagues have done much work in a series of papers [74][77], covering the fields of sensor failure, actuator sluggishness and surface lose problems. In [74], he used the NN as a quick failure detector and identifier of a typical damage to the left stabilizer of a high-performance aircraft involving a missing surface and a stuck actuator. A FDI is used to recognize the failure, and then the fault accommodation (FA) is accomplished.

Other neural network approach to reconfigurable flight control systems are also surveyed. In [23], Chiang and Youssef presented a Fault Detection, Isolation and Estimation (FDIE) scheme that uses the general regression neural network to approximate a dynamic system and detect the failure, a fuzzy isolator is also utilized to isolate failure. This FDIE scheme is applied to a F-16's short period mode of longitudinal dynamic equations which contain hard non-linearities and several parasitic dynamics. Results of the case studies presented shows the remarkable performance of the proposed FDIE scheme. Another advantage is that such a highly parallel structured neural network can be easily implemented in real-time application, and by allowing the neural network to have more neurons would result in better accuracy.

However, the reconfigurable control methods involve an extensive design to accommodate all possible control system failures at different conditions in the aircraft flight envelope. This in turn requires tremendous memory space in the on-board computer. Moreover, the approaches are based on linear techniques and are not perfectly suitable to the self-accommodation problems. To remedy these shortcomings, a restructurable flight control system is developed.

In [76], Napolitano *et al.* proposed a NN based nonlinear controller. This novel controller can bring the aircraft that has sustained extensive damage to a vital control surface, back to a new equilibrium condition. Only on-line learning is employed in the strategy; the parameters of the NN are changed in real time to accommodate the aircraft's sudden on-line damages, and there are no previous models of damage for off-line training. The learning algorithm used for this NN is also the EBP algorithm. The on-line reconstruction strategy has been numerically simulated using the software that was recently distributed for AIAA control design challenge. The types of damages considered were involving a stuck actuator at $-20°$ occurring to the left stabilator, and a reduction of 50% of the aerodynamic surface. Comparison results show that the standard BP algorithm and the conventional autopilot control systems produce unacceptable responses and both control laws could not regain control of the aircraft after the occurrence of the control surface damages, whereas the EBP algorithm for controller displays desirable performance. However, an increase in computational efforts is required by the EBP algorithm with respect to the standard BP algorithm was also reported.

Sadhukhan and Fetieh presented an exact inverse neuro-controller with full state feedback control system in [105] for the linearized longitudinal dynamics of an F8 aircraft model, with special emphasis on indirect adaptive control strategy. To obtain a decoupled response to pilot pitch rate and velocity, this paper demonstrated the controller's ability to learn the inverse dynamics during the course of flight. A linear NN was used along with off-line training and then tuning its weights on-line. Simulation studies show that the controller provided stable and near desired response in the presence of modeling uncertainty up to 30%, actuator sluggishness, and battle damage to partially missing flight control surfaces. The results presented in this paper have demonstrated the feasibility of using neuro controllers that are able to learn the vehicle dynamics on-line and to compensate for parameter uncertainties as well as actuator sluggishness. The decoupled responses of this flight control system for modeling uncertainty larger than 30% were unacceptable because the closed loop system entered into the unstable region during the transient period.

Lastly, it is worth noting that recently, some new NN based control strategies have been developed to overcome the shortcomings of the traditional reconfigurable schemes [18][10]. In [18], using a feedback linearized controller, a neural network based direct adaptive control approach is described. The scheme eliminates the need for parameter identificaiton during the recovery phase, and limits the potential need for parameter identification in the problem of control re-allocation following a failure. The test results on a X-36 aircraft demonstrates that it is a highly effective and flexible approach to control reconfiguration[18] [10].

2.3. High Angle of Attack Supermaneuver

Supermaneuverability is defined as the ability to maneuver an aircraft up to and beyond the stall angle of attack. Fighter aircrafts capable of maneuvering at extremely high angle of attack without fear of loss of control will have significant advantages over opponents without such capabilities. However, the design of flight control systems for maneuvering in this regime is complicated by the decreased effectiveness of aerodynamic control surfaces at high angle of attack and an increase in the nonlinear behavior of the aircraft.

In [101], Rokhaz and Steck has trained an appropriately constructed NN to predict the force and moment coefficients of a $70°$ sweep delta wing during a high angle of attack excursion. The angle of attack time history used for the simulation was a sinusoidal motion from $0°$ to $90°$ and returning to $0°$. The authors also explored the ability of NN to associate a stick control motion with a maneuver and to provide correct control inputs to the aircraft. Both these neural networks were constructed in the feed-foreword architecture, and the neurons in the hidden layers had the sigmoidal activation function. It has demonstrated that the NN was able to learn the model of the aerodynamic behavior of the configuration. Comparing the predicted results with those of the experiment, although certain discrepancies are present, the overall agreement between the two sets is tolerable. The second neural network trained to associate a given control stick motion with a predetermined pitch rate is providing a type of control gearing between the longitudinal stick and the thrust deflection angle. Certain discrepancies are present in the results, although the overall prediction result was quite accurate. The prediction of the thrust deflection angles does not possess sufficient accuracy to be used as control inputs to the aircraft as from the experimental results, small discrepancies in the thrust deflection angle result in large differences in the angle of attack. This is because of the sensitivity of the dynamic model to control deflections.

In these sections, we have reviewed the different neuro-flight-control strategies and aircraft flight control applications using neuro-controller. Among varied network structures, *i.e.*, MLP, MFN, RBFN, etc., in this book we are mainly concerned about the RBFN based identification and adaptive control because of their good properties for on-line or sequential adaptation, and being insensitive to the order of presentation of the signals used for adaptation. Before proceeding further, a brief review of RBFN and its applications in flight control is presented.

3. Radial Basis Function Neural Network

Originally, RBF method was used for strict multivariable interpolation, and for this the RBF model requires as many hidden neurons as data points. In [96], Powell presented a survey of the early research work on applications of RBF technique for the strict interpolation problem. The conditions for applying RBFN to multivariable interpolation are also discussed in [96]. The main conclusion from [96] is that RBFN can provide a highly promising interpolation approach to deal with irregularly positioned data points. Further, Broomhead and Lowe [12] removed the strict interpolation restriction and set up a two-layer network structure where the RBFs are employed as computation units in the hidden layer.

The basic structure of RBFN is shown in Fig.1.9. As indicated in the figure, the RBFN consists of a hidden layer and an output layer. The neurons in the hidden layer provide a set of "functions" that constitute a "basis" for the network input vectors when they are expanded into the hidden unit space. The output of the RBFN is a linear combination of the outputs from its hidden units.

If only the topologies of the networks are considered, RBFN can be viewed as a special kind of MFN. Nevertheless, considering the node characteristics and the training algorithm, RBFN is very different from multilayer feedforward network. The node characteristics for MFN are usually chosen as sigmoidal

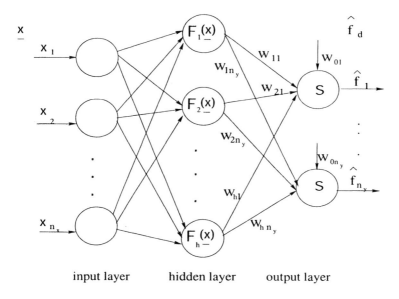

Figure 1.9. Structure of RBF neural network

functions, while for RBFN, various radial basis functions are employed. Commonly used RBF's basis functions $\phi(r)$ include,

1. Thin-plate-spline function: $\quad \phi(r) = r^2 \log(r)$

2. Multiquadric function: $\quad \phi(r) = (r^2 + \sigma^2)^{1/2}$

3. Inverse multiquadric function: $\quad \phi(r) = \frac{1}{(r^2+\sigma^2)^{1/2}}$

4. Gaussian function: $\quad \phi(r) = \exp(-r^2/\sigma^2)$

where r represents the Euclidean distance between a center μ and the input data point ξ, i.e. $r = \|\xi - \mu\|$. σ is a real variable to be decided by users for the RBFs. Among the above-mentioned RBFs, the Gaussian function is the normal choice, because it is suitable not only in generalizing a global mapping but also in refining local features without much altering the already learnt mapping. For simplicity, hereafter in this book it will be tacitly agreed that a RBFN refers to a RBFN with Gaussian function provided it is not pointed out clearly.

Mathematically, the output of an RBFN is given by,

$$\mathbf{f}(\xi) = \underline{\mathbf{w}}_0^T + \sum_{k=1}^{h} \underline{\mathbf{w}}_k^T \phi_k(\xi) \qquad \xi \in R^{n_x} \qquad (1.7)$$

$$\phi_k(\boldsymbol{\xi}) = \exp(-\frac{1}{\sigma_k^2}\|\boldsymbol{\xi} - \boldsymbol{\mu}_k\|^2) \tag{1.8}$$

where $\boldsymbol{\xi}$ is the input vector of the network, h indicates the total number of hidden neurons, $\boldsymbol{\mu}_k$ and σ_k refers to the center and width of the kth hidden neuron. $\|\cdot\|$ denotes the Euclidean norm. The function $\mathbf{f}(\cdot)$ is the output of RBFN, which represents the network approximation to the actual output. The coefficient $\underline{\mathbf{w}}_k$ is the connection weights of the ith hidden neuron to the output neurons and $\underline{\mathbf{w}}_0$ is the bias term. Note in the book, a row vector is represented by an underlined bold letter.

Compared to the traditional MFN, RBFN has the following advantages:

- Good generalization ability. The structure of the MFN tends to cause the problem of over generalization. For example, in pattern classification, MFN may be trained to have high accuracy in classifying patterns from a set of known categories, but it will also classify any pattern which is out of those categories as one of them. On the other hand, for RBF classifier, especially those based on Gaussian function, the network learns the pattern probability density instead of dividing up the pattern space, so as to avoid this problem.

- Simple topological structure. RBFN is a two-layer network whose outputs are a linear combination of the hidden layer functions. It is generally much faster to complete the training than MFN and therefore suitable for real-time on-line learning. Moreover, due to their simple topological structures, RBFNs have the ability to reveal how learning proceeds in an explicit manner [106], so they are also good candidates for hardware implementation.

- Fast convergence rate and insensitivity to the local minimum. The popular training algorithm for MFNs is Back Propagation (BP) algorithm, a kind of gradient method that was first developed by Werbos [124], Rumelhart [103]. It is known that using MFN and the BP algorithm, problems arise with local minima and saddle points, and the algorithm itself has a very slow convergence rate, especially when the system is very complex or/and has a large input-output size. In comparison to this, a range of varied training algorithms has been developed to facilitate the use of RBFN [21][68][25][43], such as Least Mean Square (LMS), Orthogonal Least Squares (OLS), Extended Kalman Filter (EKF), etc. They have been demonstrated to be effective in overcoming the shortcomings confronted by using MFNs.

Although RBFNs possess advantages when applied to the adaptive control of nonlinear systems, with an increase in the hidden neurons and the network's inputs, the computational effort increases exponentially, which limits the applications of RBFN to flight control. It was indicated from the literature survey that in most of the applications to aircraft flight control, feed-foreword network with BP learning algorithm or its extension has been used as the main paradigm. However, it is also well known that using feed-foreword network and the BP algorithm, problems arise with local minima and saddle points, and the algorithm itself has a very slow convergence rate. In this book, we are concentrated on developing new nonlinear adaptive controller using RBF network for aircraft flight applications, which produces less hidden neurons than the conventional RBFN controller.

In the last section of this chapter, we briefly survey the current status in using the RBFN controller for aircraft flight applications, summarize the problems existing in the controller design and discuss the possible solutions for these problems.

4. Aircraft Flight Control Applications Using RBF Network

Neural network for flight control was also studied in [65], where McGrane *et al.* present the application of two neural network control paradigms, namely recurrent network controller and spatially localized network controller, to stabilize the longitudinal axis dynamics of a highly nonlinear aircraft, to adapt to gradual variations in plant parameters, and to adequately track the reference model over a range of flight conditions. The ability of each controller to adapt its performance to variations in the plant dynamics is assessed, as well as the ability of the controllers to generalize to situations for which they were not trained. The results obtained show that the spatially localized network controller (an alternate name for RBF method) can provide a better tracking performance and exhibits the potential for on-line application of learning control systems.

As a spatially localized network, since the early nineties, RBFNs have found wide applications in nonlinear control, such as robot control [27][28], process control [9][64], etc. However, in aircraft flight control, there are only limited papers which explore the application of RBFN controller.

In [113], Singh *et al.* used a RBFN to suppress the wing rock for a slender delta wing configuration. Calise presented the use of RBFN to capture variations in Mach number in [15], as these variations are difficult to be represented by polynomial functions in the transonic region. Calise also proposed an on-line adaptive architecture that employs a neural network to compensate for inversion error presented when feedback linearization methods are used to control a dynamic process [16]. A stable weight adjustment rule is derived and utilized by the neural network controller in the paper. An aircraft application is used to illustrate the benefits of this scheme for nonlinear dynamics in both the states and the controls. In using the on-line architecture in combination with the on-line algorithm for weight adjustment, the results obtained show good tracking ability and desirable transient characteristics. A comparison study was carried out with the off-line trained neural network architecture, where the control becomes unstable in situations when the aircraft is flown outside the envelope used in the off-line training. The on-line adaptive networks used in this paper has the advantage in demonstrating rapid adaptation and ability to correct for inversion errors induced by modeling uncertainties and partial loss of actuators effectiveness.

However, in most of the control applications presented in the literature, the RBFN's structure is implemented by a classical approach. That is, the number of hidden neurons used in the RBFN is determined according to prior knowledge, and it assumes that the centers and widths of the Gaussian functions are known *a priori*. Then, only the weights connecting the hidden neurons and output neurons are updated. In [2], Polycarpou and Youssef investigated such an implementation by using the RBFN as an on-line approximator of the aircraft pitch dynamics, combined with a nonlinear control law to improve the closed-loop system performance. However, in practice, designers are often puzzled by the problems of choosing the right number of hidden units, and it is very difficult to determine the exact value for the parameters such as centers, widths of the hidden neurons. The use of the inaccurate centers and widths, usually results in a deterioration of the performance especially in the time-varying system. In order to improve the performance, a large number of hidden neurons has to be selected, which in turn results in a slow on-line learning and may cause the problem of overfitting [8][122].

Research efforts have been undertaken to solve these problems. In a series of papers [19][21] [25], Billings and his coworkers have worked out an unsupervised learning scheme to accurately estimate the centers of the RBF. However, they have used the thin-plate-spline function instead of the Gaussian function, where no widths are to be considered. Contrary to the conventional approaches of only tuning the weights, recently fully tuned RBFN begin to exhibit their great potential for accurate approximation and identification. In a fully tuned RBFN, not only the weights of the output layer, but also the other parameters of the network are updated, so that the nonlinearities of the dynamic system can be captured as quickly as possible [8][91].

In determining the proper number of hidden neurons for a given problem, [48] introduced the concept of building up of the hidden neurons with the update of the RBF parameters being done by a gradient descent algorithm. An alternative approach is to start with as many as hidden units as the number of inputs and then reduce them using a clustering algorithm which essentially puts patterns that are close in the input space into a cluster to remove the unnecessary hidden neurons [72]. However, the main learning scheme for these studies is of batch type, which is not suitable for online learning. In [94], Platt proposed a sequential learning algorithm for a fully tuned RBFN to remedy this drawbacks. In Platt's Resource Allocating Network (RAN) algorithm, hidden neurons are added based on the novelty of the new data and the parameters of the network (weights, as well as widths and centers) are estimated using a LMS method. Platt showed the resulting network topology to be more parsimonious than the classical RBF networks. Kadirkamanathan and Niranjan [44] proposed modifications to improve the RAN algorithm by using an extended Kalman filter (EKF) instead of the LMS to estimate the network parameters. The resulting network called RANEKF is more compact and has better accuracy than RAN. A further improvement to RAN and RANEKF was proposed by [60] in which a pruning strategy was introduced to remove those neurons that consistently made little contributions to the network output. The resulting network, called Minimal RAN (MRAN), was shown to be more compact than and RANEKF for several applications in the areas of function approximation and pattern classification [60][59]. Survey papers about the pruning strategy can be found in [99][3].

Although for nonlinear system identification, various algorithms for fully tuned RBFN with growing/pruning strategy has been developed, such as MRAN, ONSAHL [61][43] etc., the control applications for these algorithm is still not available. The main difficulty that exists in applying NNs in real-time control is that their learning time is excessive in comparison to the time period that characterize the evolution of the process dynamics. A second issue of importance in real-time control applications is that, whatever adaptation algorithm is ultimately employed to adjust the parameters in a neural network, it must ensure stability of the controlled process. Therefore, the control structures designed and the parameter tuning rules adopted must meet the requirement of the stability and convergency for the overall system.

It is in this context that we investigate the nonlinear adaptive control scheme using a fully tuned RBF neural networks, to improve the speed and accuracy of on-line identification and control. In Part I of this book, the indirect adaptive neuro-control strategies are studied. A stable identification scheme using a fully tuned RBFN is proposed [53]. The performance of other sequential learning algorithms for a fully tuned RBFN, such as MRAN, ONSAHL, are also evaluated [118][43][102]. Moreover, a novel algorithm named Extended MRAN (EMRAN), which combines the MRAN with a 'winner neuron' strategy is proposed for fast on-line learning [54]. Then these identification strategies are incorporated into the indirect adaptive control scheme and their performances are evaluated.

Further in Part II of the book, the NN based direct control strategies are studied to ensure the stability of the controlled process, with the aim of finding the proper control structure for the aircraft flight system. We extend the existing schemes by deriving tuning rules for a fully adaptable RBFN using Lyapunov stability theory [55]. The proposed tuning rule not only ensures the stability of the overall system, but also greatly improves the tracking accuracy. Moreover, the robustness of this control scheme and the proposed tuning rule is analyzed in the presence of the approximation error and model errors. Simulation studies based on high performance aircraft models demonstrate the theoretical results presented.

I

NONLINEAR SYSTEM IDENTIFICATION AND INDIRECT ADAPTIVE CONTROL SCHEMES

In this part, different learning algorithms for the fully tuned RBFN are introduced and analyzed for nonlinear system identification, including some newly developed algorithms with growing and pruning capabilities. Using these identification schemes, indirect adaptive control strategies are studied in detail. The performance of an indirect neuro-controller is evaluated for aircraft flight control based on a linearized longitudinal F8 aircraft model.

Part I is divided into three chapters. A stable identification scheme using the fully tuned RBFN is developed in Chapter 3. The tuning rule for updating all the parameters of the network is derived using the Lyapunov synthesis approach. This extends the existing algorithms for only tuning the weights of the RBFN and guarantees the stability of the overall system. The proposed method is then utilized to identify several nonlinear systems, including a complicated time-varying nonlinear missile dynamics. Compared with the RBFN with only weights tuning, the proposed algorithm can identify the nonlinear system with improved accuracy.

In Chapter 4, the performance of the newly proposed MRAN and ONSAHL algorithm (both based on fully tuned RBFN) is investigated for nonlinear system identification. By incorporating a "winner neuron" strategy, a new algorithm with the name of Extended MRAN (EMRAN) is proposed to improve the computational speed of the MRAN algorithm. The performance of the EMRAN is analyzed and compared with that of the MRAN using benchmark problems.

Based on the algorithms developed for nonlinear system identification, *i.e.*, Lyapunov based fully tuned RBFN, MRAN, etc., Chapter 5 explores the indirect adaptive control scheme, with aircraft flight control applications. First the indirect adaptive control scheme with off-line training and on-line control is investigated. Simulation studies reveal that in the off-line training stage, with GAP strategy, a more "economical" network is implemented. This RBFN is then utilized as the initial network in the control stage. It is demonstrated from the comparison study that to get better performance, each training sample should be randomly arranged to form the training data pair. In on-line control mode, the performance of different neuro-controllers, including RBFN with fixed number of hidden neurons, Growing RBFN and GAP RBFN are evaluated based on a linearized F8 aircraft model. When the number of hidden neurons are fixed or grow automatically, good tracking results are observed. However, the performance deteriorates, or even diverges if a pruning strategy is incorporated. Simulation studies also indicate a similar performance using different tuning rules such as extended Kalman filter (EKF) in MRAN, or the Lyapunov-based stable tuning rule.

Since an off-line training is required, the off-line training and on-line control is not practical when the system dynamics is suddenly changed due to unexpected situations such as actuator sluggishness. Therefore the case of on-line learning/on-line control scheme is studied. However, simulation results show that without the prior knowledge of the system, divergence is observed in this control scheme. To remedy this problem, further research has been carried out using direct adaptive control strategies in the latter part of the book.

Chapter 2

NONLINEAR SYSTEM IDENTIFICATION USING LYAPUNOV-BASED FULLY TUNED RBFN

1. Introduction

In recent years, nonlinear system identification using neural network has become a widely studied area because of its close relationship to the system control. Basically, any good identification scheme that incorporates RBFN should satisfy two criteria: (i) The parameters of the RBFN are tuned appropriately so that its output to an input signal can approximate the response of the real system to the same input with good accuracy. (ii) The network structure is compact and the parameter adaptive law is efficient so that fast on-line learning can be implemented.

The parameters for the RBFN with Gaussian function consist of weights connecting the hidden layer and output layer, widths, centers of the Gaussian functions. The classical approach to the network implementation is to fix the number of hidden neurons *a priori* along with its centers and widths, based on some properties of the input data, and then estimate the weights using algorithm such as BP, LMS, etc. [19][68][20][67]. This method has two disadvantages. First, it is difficult to select the exact network structure; Second, with the option of only tuning the weights, the performance is poor with the use of inaccurate centers and widths.

To determine the proper number of hidden neurons for a given problem, and overcome the drawbacks caused by only tuning the weights of the RBFN, Platt [94] proposed Resource Allocating Network (RAN) for a fully tuned RBFN, where hidden neurons are added automatically provided some criteria are satisfied. Kadirkamanathan and Niranjan improve the RAN by using EKF to estimate the network parameters [44]. Further, Lu *et al.* proposed a Minimal RAN algorithm in which a pruning strategy was introduced to remove the neurons that consistently made little contributions to the network outputs [60].

However, all these algorithms, including RAN and its extensions, mainly address the identification of nonlinear systems without external inputs, and hence stability was not an issue. When they are applied to identify systems with external inputs, problems such as the stability of the overall scheme and convergence of the approximation error arise.

To identify systems with external inputs, Ni *et al.* used two fixed RBF networks to mimic a nonlinear affine system [88]. The tuning rule for updating the weights connecting the hidden layer and output layer was derived based on Lyapunov synthesis approach, which ensured the stability of the overall system. Liu and Kadirkamanathan [58] also derived the weight tuning law independently using a similar strategy. Their derivation is based on a general nonlinear dynamic system model, and to implement a compact network structure, a Growing RBF Network (GRBFN) was used.

Unfortunately, both the [88] and [58] used the classical weights-tuning algorithms, which are based on the assumption that the hidden units' centers and widths are selected properly *a priori*. In this chapter, we extend Liu's algorithm by developing a scheme for a fully adaptable GRBF network, to realize more effective on-line identification. Using Lyapunov stability theory, the parameter tuning law for this fully tuned GRBF network is derived that guarantees the stability of the overall system. To demonstrate the efficiency of the proposed scheme, the method is applied for on-line identification of a complex time-varying nonlinear MIMO missile dynamics. Further, simulation studies confirm that stability and reduction in approximation error can also be extended to a growing and pruning (GAP) RBF network, as suggested in [61], where those neurons that consistently make little contribution to the network outputs are removed.

[**Notation**] In this book, vectors are represented by bold lower case letters (for example, **e**), matrices are indicated by bold upper case letters, e.g. **E**, and scalars are represented by ordinary letters (not bold), for example, E or e.

2. Stable Identification Using Lyapunov-Based Fully Tuned RBF Network

2.1. Identification Strategy and System Error Dynamics

Fig.2.1 describes the proposed stable identification scheme, where the multivariable nonlinear dynamic system with states and inputs is given by,

$$\Sigma: \quad \dot{\mathbf{x}}(t) = \mathbf{f}(\mathbf{x}(t), \mathbf{u}(t)) \quad \mathbf{x}(0) = \mathbf{0} \quad (2.1)$$

where **x** is $n_y \times 1$ state vector in the state space $\overline{\mathbf{X}}$, **u** is $n_u \times 1$ input vector in the input space $\overline{\mathbf{U}}$. It is assumed that $\overline{\mathbf{X}}$ and $\overline{\mathbf{U}}$ are compact sets and $\mathbf{f}()$ is smooth. By subtracting and adding $\mathbf{Ax}$, where $\mathbf{A} \in R^{n_y \times n_y}$ is a Hurwitz

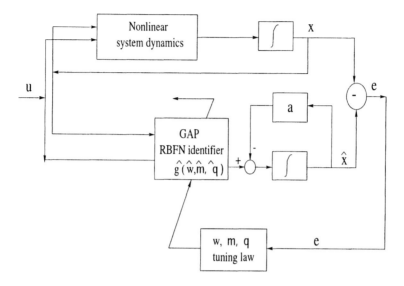

Figure 2.1. The stable sequential identification framework

matrix, Σ becomes,

$$\Sigma: \quad \dot{\mathbf{x}}(t) = \mathbf{A}\mathbf{x}(t) + \mathbf{g}(\mathbf{x}(t), \mathbf{u}(t)) \quad (2.2)$$

where $\mathbf{g}(\mathbf{x}(t), \mathbf{u}(t)) = \mathbf{f}(\mathbf{x}(t), \mathbf{u}(t)) - \mathbf{A}\mathbf{x}(t)$ describes the system nonlinearity. In practice, apart from being Hurwitz, $\mathbf{A}$ can be selected to represent the desired system dynamics having good response characteristics like damping, overshoot, etc.

Setting the RBF network's inputs $\boldsymbol{\xi}$ as $\boldsymbol{\xi} = [\mathbf{x}(t)^T, \mathbf{u}(t)^T]^T$, the problem of system identification can be converted into a nonlinear function approximation problem (For detailed description of RBFN, refer to Section 2.3). To avoid pre-determining the number of hidden neurons in the network and also with the aim of implementing a compact network structure, a GRBF network is selected to approximate the nonlinear input-output mapping $\mathbf{g}(\mathbf{x}(t), \mathbf{u}(t))$. The GRBF network based system model can be written as,

$$\begin{aligned}\dot{\mathbf{x}}(t) &= \mathbf{A}\mathbf{x}(t) + \sum_{k=1}^{h} \mathbf{w}_k^{*T} \exp\left(-\frac{1}{\sigma_k^{*2}} \|\boldsymbol{\xi} - \boldsymbol{\mu}_k^*\|^2\right) + \epsilon_h \\ &= \mathbf{A}\mathbf{x}(t) + \mathbf{W}^{*T} \boldsymbol{\phi}(\boldsymbol{\mu}^*, \sigma^*, \boldsymbol{\xi}) + \epsilon_h \end{aligned} \quad (2.3)$$

where $\mathbf{W}^*$ is $h \times n_y$ optimal weight matrix (the variable h indicates the number of hidden neurons), $\underline{\mathbf{w}}_k^*$ is the kth row vector of $\mathbf{W}^*$, and ϕ is $h \times 1$ Gaussian function vector, which is determined by the optimal centers μ^* and widths σ^*. For simplicity ϕ^* will be used instead of $\phi(\mu^*, \sigma^*, \xi)$, hereafter. The approximation error ϵ_h is defined as,

$$\epsilon_h = g(x, u) - W^{*T}\phi^* \tag{2.4}$$

Using approximation theory, ϵ_h can be reduced arbitrarily by increasing the number of h [93]. Then, it is reasonable to assume that ϵ_h is bounded by a constant ϵ_H, and

$$\epsilon_H = \sup_{x \in \overline{X}, u \in \overline{U}} \|\epsilon_h(x, u)\| \tag{2.5}$$

For on-line identification, the above nonlinear function $g()$ can be approximated by a GRBF network as,

$$\hat{g}(x, u) = \sum_{k=1}^{h} \underline{\hat{w}}_k^T \exp(-\frac{1}{\hat{\sigma}_k^2} \|\xi - \hat{\mu}_k\|^2) = \hat{W}^T \phi(\hat{\mu}, \hat{\sigma}, \xi) \tag{2.6}$$

where $\hat{W}$ is the estimated weight matrix, and $\underline{\hat{w}}_k$, the estimated weights, is the kth row vector of $\hat{W}$. $\hat{\sigma}_k$ and $\hat{\mu}_k$ are the estimated center and width for the kth hidden neuron. $\|\cdot\|$ denotes the Euclidean norm. $\hat{\phi} = \phi(\hat{\mu}, \hat{\sigma}, \xi)$ is the corresponding vector expression of the hidden neurons.

Substituting $\hat{g}()$ into Eq.(2.2), from Fig.2.1, the identification model is,

$$\dot{\hat{x}}(t) = A\hat{x}(t) + \hat{W}^T\hat{\phi} \tag{2.7}$$

Defining the approximation error as $e = x - \hat{x}$, from Eq.(2.2) and Eq.(2.7), the error dynamics of the overall system can be expressed as,

$$\begin{aligned}
\dot{e} &= Ae + W^{*T}\phi^* - \hat{W}^T\hat{\phi} + \epsilon_h \\
&= Ae + (\hat{W} + \tilde{W})^T(\hat{\phi} + \tilde{\phi}) - \hat{W}^T\hat{\phi} + \epsilon_h \\
&= Ae + (\hat{W}^T\tilde{\phi} + \tilde{W}^T\hat{\phi} + \tilde{W}^T\tilde{\phi}) + \epsilon_h \\
&\approx Ae + (\hat{W}^T\tilde{\phi} + \tilde{W}^T\hat{\phi}) + \epsilon_h
\end{aligned} \tag{2.8}$$

where $\tilde{W}$ is the difference between W^* and $\hat{W}$, i.e. $\tilde{W} = W^* - \hat{W}$, $\tilde{\phi}$ is the difference between ϕ^* and $\hat{\phi}$, i.e. $\tilde{\phi} = \phi^* - \hat{\phi}$, and the higher order item $\tilde{W}^T\tilde{\phi}$ has been neglected.

2.2. Stable Parameter Tuning Rules

To derive the stable tuning laws, choose the following Lyapunov candidate function,

$$V = \frac{1}{2}e^T P e + \frac{1}{2}tr(\tilde{W}^T\tilde{W}) + \frac{1}{2}\tilde{\phi}^T\tilde{\phi} \tag{2.9}$$

where $\mathbf{P}$ is an $n_y \times n_y$ symmetric positive definite matrix. The derivative of the Lyapunov function is given by,

$$\dot{V} = \frac{1}{2}(\mathbf{e}^T\mathbf{P}\dot{\mathbf{e}} + \dot{\mathbf{e}}^T\mathbf{P}\mathbf{e}) + tr(\tilde{\mathbf{W}}^T\dot{\tilde{\mathbf{W}}}) + \tilde{\boldsymbol{\phi}}^T\dot{\tilde{\boldsymbol{\phi}}} \qquad (2.10)$$

substituting Eq.2.7 into Eq.2.10,

$$\begin{aligned}\dot{V} &= -\frac{1}{2}\mathbf{e}^T\mathbf{Q}\mathbf{e} + \boldsymbol{\epsilon}_h^T\mathbf{P}\mathbf{e} + \tilde{\boldsymbol{\phi}}^T\hat{\mathbf{W}}\mathbf{P}\mathbf{e} \\ &+ \hat{\boldsymbol{\phi}}^T\tilde{\mathbf{W}}\mathbf{P}\mathbf{e} + tr(\tilde{\mathbf{W}}^T\dot{\tilde{\mathbf{W}}}) + \tilde{\boldsymbol{\phi}}^T\dot{\tilde{\boldsymbol{\phi}}}\end{aligned} \qquad (2.11)$$

where $\mathbf{Q} = -(\mathbf{PA} + \mathbf{A}^T\mathbf{P})$. Since $\mathbf{A}$ is Hurwitz, a Lyapunov function can always be found that has a unique solution when a symmetric positive definite matrix, $\mathbf{Q}$, is given. Moreover, in Eq.2.10,

$$\begin{aligned}tr(\tilde{\mathbf{W}}^T\dot{\tilde{\mathbf{W}}}) &= \sum_{i=1}^{n_y}\tilde{\mathbf{w}}_i^T\dot{\tilde{\mathbf{w}}}_i \\ \hat{\boldsymbol{\phi}}^T\tilde{\mathbf{W}}\mathbf{P}\mathbf{e} &= \sum_{i=1}^{n_y}\hat{\boldsymbol{\phi}}^T\tilde{\mathbf{w}}_i(\mathbf{P}\mathbf{e})_i = \sum_{i=1}^{n_y}\tilde{\mathbf{w}}_i^T\hat{\boldsymbol{\phi}}(\mathbf{P}\mathbf{e})_i\end{aligned} \qquad (2.12)$$

Eq.2.10 becomes,

$$\begin{aligned}\dot{V} &= -\frac{1}{2}\mathbf{e}^T\mathbf{Q}\mathbf{e} + \boldsymbol{\epsilon}_h^T\mathbf{P}\mathbf{e} + \tilde{\boldsymbol{\phi}}^T(\hat{\mathbf{W}}\mathbf{P}\mathbf{e} + \dot{\tilde{\boldsymbol{\phi}}}) \\ &+ \sum_{i=1}^{n_y}(\tilde{\mathbf{w}}_i^T\hat{\boldsymbol{\phi}}(\mathbf{P}\mathbf{e})_i + \tilde{\mathbf{w}}_i^T\dot{\tilde{\mathbf{w}}}_i)\end{aligned} \qquad (2.13)$$

where $\dot{\tilde{\mathbf{w}}}_i$ and $\tilde{\mathbf{w}}_i$ are the ith column vector of matrix $\dot{\tilde{\mathbf{W}}}$ and $\tilde{\mathbf{W}}$ respectively, and $(\mathbf{P}\mathbf{e})_i$ represents the ith element of the vector $\mathbf{P}\mathbf{e}$. Therefore, letting $\dot{\tilde{\mathbf{w}}}_i$ and $\dot{\tilde{\boldsymbol{\phi}}}$ be expressed as,

$$\begin{aligned}\dot{\tilde{\mathbf{w}}}_i &= -\hat{\boldsymbol{\phi}}(\mathbf{P}\mathbf{e})_i, \quad i = 1,...,n_y \\ \dot{\tilde{\boldsymbol{\phi}}} &= -\hat{\mathbf{W}}\mathbf{P}\mathbf{e}\end{aligned} \qquad (2.14)$$

Eq.2.12 will become,

$$\dot{V} = -\frac{1}{2}\mathbf{e}^T\mathbf{Q}\mathbf{e} + \boldsymbol{\epsilon}_h^T\mathbf{P}\mathbf{e} \qquad (2.15)$$

and Eq.2.15 satisfies,

$$\dot{V} \leq -\frac{1}{2}\|\mathbf{e}\|\lambda_{min}(\mathbf{Q})\|\mathbf{e}\| + \epsilon_H\|\mathbf{P}\|\|\mathbf{e}\| \qquad (2.16)$$

From Eq.2.16, $\dot{V}$ is negative if,

$$\|\mathbf{e}\| > \frac{2\epsilon_H \|\mathbf{P}\|}{|\lambda_{min}(\mathbf{Q})|} = E_a \qquad (2.17)$$

Using the universal approximation proposition [93], by increasing the number h, ϵ_H can be reduced arbitrarily small, which means that E_a tends to zero when $h \to \infty$ and the negativeness of the Lyapunov function can be guaranteed. However, it should be noted that in the real implementation, $h \to \infty$ is impossible, in this case the uniform ultimate boundedness (UUB) of the error signals is achieved. Since $\dot{\tilde{\mathbf{w}}}_i = \dot{\mathbf{w}}_i^* - \dot{\hat{\mathbf{w}}}_i$, $\dot{\tilde{\phi}} = \dot{\phi}^* - \dot{\hat{\phi}}$ and $\dot{\mathbf{w}}_i^* = \mathbf{0}$, $\dot{\phi}^* = \mathbf{0}$, the tuning rules in Eq.2.13 become,

$$\dot{\hat{\mathbf{w}}}_i = \hat{\phi}(\mathbf{Pe})_i, \quad i = 1, ..., n_y \qquad (2.18)$$

$$\dot{\hat{\phi}} = \hat{\mathbf{W}}\mathbf{Pe} \qquad (2.19)$$

2.3. Implementation of the Tuning Rule

In the above tuning rules, the Gaussian function $\hat{\phi}$ is embedded with the centers' locations $\hat{\mu}$ and widths $\hat{\sigma}$, i.e. $\hat{\phi} = \phi(\hat{\mu}, \hat{\sigma}, \xi)$. Combining all the parameters adaptable into a big vector, $\chi = [\hat{\mathbf{w}}_1, \hat{\mu}_1^T, \hat{\sigma}_1, \cdots, \hat{\mathbf{w}}_h, \hat{\mu}_h^T, \hat{\sigma}_h]^T$, a simple updating rule for χ will be derived in this section. Because the real implementation is carried out in a discrete-time framework, the updating laws is derived under discrete form.

First, Eq.2.18 can be converted into,

$$\dot{\hat{\mathbf{w}}}_i = \hat{\phi}(\mathbf{Pe})_i, \quad i = 1, ..., n_y$$
$$\Rightarrow \dot{\hat{\mathbf{W}}}^T = \mathbf{Pe}\hat{\phi}^T \Rightarrow \dot{\underline{\hat{\mathbf{w}}}}_k^T = \mathbf{Pe}\hat{\phi}_k, \quad k = 1, ..., h \qquad (2.20)$$

where $\dot{\hat{\mathbf{w}}}_i$ is the ith column and $\underline{\hat{\mathbf{w}}}_k$ is the kth row vector of matrix $\hat{\mathbf{W}}$ respectively. Because $\hat{\phi}_k$ is in fact the derivative of $\hat{g}()$ to the weights $\underline{\hat{\mathbf{w}}}_k$, this equation can be converted into a discrete form,

$$\underline{\hat{\mathbf{w}}}_k^T(n+1) = \underline{\hat{\mathbf{w}}}_k^T(n) + \tau \frac{\partial \hat{\mathbf{g}}}{\partial \underline{\hat{\mathbf{w}}}_k^T}\mathbf{Pe}(n) \quad k = 1, ..., h \qquad (2.21)$$

where τ is the sampling time.

To determine the tuning rules for the centers and width, from Eq.2.19,

$$\dot{\hat{\phi}}_k = \underline{\mathbf{w}}_k \mathbf{Pe} \quad k = 1, \cdots, h \qquad (2.22)$$

At the same time, since $\hat{\phi}_k$ is a Gaussian function, i.e.,

$$\hat{\phi}_k = \exp\{-\frac{\|\xi - \hat{\mu}_k\|^2}{\hat{\sigma}_k^2}\} \qquad (2.23)$$

for a given input $\boldsymbol{\xi}$, its derivative can be expressed as,

$$\dot{\hat{\phi}}_k = \frac{\partial \hat{\phi}_k}{\partial \hat{\boldsymbol{\mu}}_k^T}\dot{\hat{\boldsymbol{\mu}}}_k + \frac{\partial \hat{\phi}_k}{\partial \hat{\sigma}_k}\dot{\hat{\sigma}}_k \qquad (2.24)$$

Note in Eq.2.22, $\underline{\mathbf{w}}_k\mathbf{P}\mathbf{e}$ is a scalar, which reflects the varying direction of $\dot{\hat{\phi}}_k$. In order to obtain the tuning rules of the center $\hat{\boldsymbol{\mu}}_k$ and width $\hat{\sigma}_k$, Eq.2.24 is partitioned into two parts:

$$\frac{\partial \hat{\phi}_k}{\partial \hat{\boldsymbol{\mu}}_k^T}\dot{\hat{\boldsymbol{\mu}}}_k = \lambda \underline{\mathbf{w}}_k\mathbf{P}\mathbf{e} \qquad (2.25)$$

$$\frac{\partial \hat{\phi}_k}{\partial \hat{\sigma}_k}\dot{\hat{\sigma}}_k = (1-\lambda)\underline{\mathbf{w}}_k\mathbf{P}\mathbf{e} \qquad (2.26)$$

where $0 < \lambda < 1$. Denote $\mathbf{r}_k = (\frac{\partial \hat{\phi}_k}{\partial \hat{\boldsymbol{\mu}}_k^T})^T = \frac{\partial \hat{\phi}_k}{\partial \hat{\boldsymbol{\mu}}_k}$, which is a column vector with $n_y + n_u$ elements. Assuming $r_{k,1}, ..., r_{k,n_y+n_u}$ are not zero, and denoting

$$\mathbf{r}_k = \begin{bmatrix} r_{k,1} \\ r_{k,2} \\ \vdots \\ r_{k,n_y+n_u} \end{bmatrix}$$

and

$$\mathbf{r}_k^{(-1)} = \begin{bmatrix} 1/r_{k,1} \\ 1/r_{k,2} \\ \vdots \\ 1/r_{k,n_y+n_u} \end{bmatrix}, \quad \mathbf{r}_k^{(-2)} = \begin{bmatrix} 1/r_{k,1}^2 \\ 1/r_{k,2}^2 \\ \vdots \\ 1/r_{k,n_y+n_u}^2 \end{bmatrix} \qquad (2.27)$$

the tuning rule of the center $\hat{\boldsymbol{\mu}}_k$ can be selected as one of the solutions for Eq.2.25,

$$\dot{\hat{\boldsymbol{\mu}}}_k = \frac{\lambda}{n_y + n_u}\mathbf{r}_k^{(-1)}\underline{\mathbf{w}}_k\mathbf{P}\mathbf{e}$$

$$= \frac{\lambda}{n_y + n_u}(\mathbf{r}_k^{(-2)}.*\frac{\partial \hat{\phi}_k}{\partial \hat{\boldsymbol{\mu}}_k})\underline{\mathbf{w}}_k\mathbf{P}\mathbf{e} \qquad (2.28)$$

and the tuning rule for the width can be derived from Eq.2.26,

$$\dot{\hat{\sigma}}_k = (1-\lambda)(\frac{\partial \hat{\phi}_k}{\partial \hat{\sigma}_k})^{-1}\underline{\mathbf{w}}_k\mathbf{P}\mathbf{e}$$

$$= (1-\lambda)(\frac{\partial \hat{\phi}_k}{\partial \hat{\sigma}_k})^{-2}\frac{\partial \hat{\phi}_k}{\partial \hat{\sigma}_k}\underline{\mathbf{w}}_k\mathbf{P}\mathbf{e} \qquad (2.29)$$

".*" is array multiplication denoting the element-by-element multiplication of two vectors. Let $\alpha = \frac{\lambda}{n_y+n_u}\mathbf{r}_k^{(-2)}$ and $\beta = (1-\lambda)(\frac{\partial \hat{\phi}_k}{\partial \hat{\sigma}_k})^{-2}$, we have,

$$\dot{\hat{\mu}}_k = (\alpha .* \frac{\partial \hat{\phi}_k}{\partial \hat{\mu}_k})\mathbf{w}_k \mathbf{Pe} \qquad (2.30)$$

$$\dot{\hat{\sigma}}_k = \beta \frac{\partial \hat{\phi}_k}{\partial \hat{\sigma}_k}\mathbf{w}_k \mathbf{Pe} \qquad (2.31)$$

Noting the elements of α as well as β in the above equations are positive, then the above two equations can be converted into the discrete form as follows,

$$\begin{aligned}\hat{\mu}_k(n+1) &= \hat{\mu}_k(n) + \tau(\alpha .* \frac{\partial \hat{\phi}_k}{\partial \hat{\mu}_k})\mathbf{w}_k \mathbf{Pe}(n) \\ &\approx \hat{\mu}_k(n) + \eta_1 \frac{\partial \hat{\phi}_k}{\partial \hat{\mu}_k}\mathbf{w}_k \mathbf{Pe}(n) \qquad (2.32)\end{aligned}$$

$$\begin{aligned}\hat{\sigma}_k(n+1) &= \hat{\sigma}_k(n) + \tau\beta \frac{\partial \hat{\phi}_k}{\partial \hat{\sigma}_k}\mathbf{w}_k \mathbf{Pe}(n) \\ &\approx \hat{\sigma}_k(n) + \eta_2 \frac{\partial \hat{\phi}_k}{\partial \hat{\sigma}_k}\mathbf{w}_k \mathbf{Pe}(n) \qquad (2.33)\end{aligned}$$

In the above equations, η_1 and η_2 are positive scalars to be selected properly *a priori* by trial and error. Integrating Eq.2.21, Eq.2.32 and Eq.2.33 by using the vector χ, the tuning rule is,

$$\chi(n+1) = \chi(n) + \eta \mathbf{\Pi}(n)\mathbf{Pe}(n) \qquad (2.34)$$

where η is the learning rate and $\eta < min(\tau, \eta_1, \eta_2)$. $\mathbf{\Pi}(n) = \nabla_\chi \hat{\mathbf{g}}(\boldsymbol{\xi}_n)$ is the gradient of the function $\hat{\mathbf{g}}()$ with respect to the parameter vector χ evaluated at $\chi(n)$, and

$$\mathbf{\Pi}(n) = [\hat{\phi}_1 I_{n_y \times n_y}; \hat{\phi}_1 \frac{2(\boldsymbol{\xi}_n - \hat{\boldsymbol{\mu}}_1)\hat{\mathbf{w}}_1}{\hat{\sigma}_1^2}; \hat{\phi}_1 \frac{2\hat{\mathbf{w}}_1}{\hat{\sigma}_1^3}\|\boldsymbol{\xi}_n - \hat{\boldsymbol{\mu}}_1\|^2; \cdots ;$$
$$\hat{\phi}_h I_{n_y \times n_y}; \hat{\phi}_h \frac{2(\boldsymbol{\xi}_n - \hat{\boldsymbol{\mu}}_h)\hat{\mathbf{w}}_h}{\hat{\sigma}_h^2}; \hat{\phi}_h \frac{2\hat{\mathbf{w}}_h}{\hat{\sigma}_h^3}\|\boldsymbol{\xi}_n - \hat{\boldsymbol{\mu}}_h\|^2] \qquad (2.35)$$

2.4. GRBF Network and Dead Zone Design

Since a GRBF network is utilized in this scheme and the network is growing from zero hidden unit, the approximation error E_a is quite large at the beginning. If $\|e\| < E_a$, then it is possible that $\dot{V} > 0$, which implies that the parameters of the network may drift to infinity. To avoid this problem and ensure the convergence of approximation error, a dead zone is incorporated in the tuning rules.

The dead zone is defined by putting a limit on the identification error and also on the magnitude of network weights. Using similar arguments as in [58], the enhanced tuning law can be written as,

$$\chi(n+1) = \begin{cases} \chi(n) + \eta \mathbf{\Pi}(n)\mathbf{P}\mathbf{e}(n) & \text{if } \|\mathbf{e}\| \geq e_0 \\ & \text{and } \|\mathbf{W}\| \leq h^{\frac{1}{2}} Z \\ \chi(n) & \text{otherwise} \end{cases} \quad (2.36)$$

In Eq.(2.36), Z is a positive scalar and $h^{\frac{1}{2}} Z$ is an upper bound on $\|\mathbf{W}\|$ (Euclidean norm of weight matrix), e_0 is selected as the required accuracy on the state error $\mathbf{e}$. According to Eq.(2.17), the size of the dead zone should be set to E_a, so that if the error $\|\mathbf{e}\|_2 < E_a$, the tuning is stopped and the parameters will not drift away. Unfortunately, since ϵ_H is difficult to be known, the exact value for E_a is unknown, therefore e_0 is used in the place of $E_a(t)$. If $e_0 > E_a$, then $\dot{V}$ is always non-positive. If $e_0 < E_a$, when the error $\|\mathbf{e}\|_2$ converges to $e_0 < \|\mathbf{e}\|_2 < E_a$, $\dot{V}$ may be positive and in this case the upper bound $h^{\frac{1}{2}} Z$ is used to prevent the parameter from drifting away. Using a GRBFN, as the network is gradually adding hidden neurons through on-line learning, E_a shrinks into a very small region inside e_0. A small dead zone (e_0) means the approximation can be more accurate, at a cost of more hidden units being added.

The growing strategy is based on the principle that if the approximation accuracy is not satisfied according to some criteria, a new hidden neuron is added [94]. In this scheme, the one proposed in Lu *et al.*'s MRAN algorithm [61] is used (For detailed description of MRAN algorithm, refer to Chapter 4): a hidden neuron is recruited when all the following three criteria are satisfied. (a) $\|\mathbf{e}(n)\| = \|\mathbf{x}(n) - \hat{\mathbf{x}}(n)\| > E_1(E_{max})$, (b) $d(n) = \|\hat{\boldsymbol{\xi}}(n) - \hat{\boldsymbol{\mu}}_r(n)\| > E_2$, (c) $e_{rmsn}(n) = \sqrt{\sum_{i=n-(N_w-1)}^{n} \frac{\|\mathbf{e}(i)\|^2}{N_w}} > E_3$, where E_1, E_2, E_3 are thresholds to be selected *a priori*, $\hat{\boldsymbol{\mu}}_r(n)$ is the center of the hidden neuron which has the closest distance to the current network input $\boldsymbol{\xi}(n) = [\mathbf{x}^T(n), \mathbf{u}^T(n)]^T$. In (c), for each observation, a sliding data window N_w is employed to include a certain number of output errors. Moreover, the additional two criteria(b)(c) are applied to ensure that the neuron to be added is sufficiently far away from all the existing neurons and the network has met the required sum squared error for the past N_w consecutive outputs. Once the new hidden unit is recruited, the associated parameters are given as: $\hat{\boldsymbol{\mu}}_{h+1} = \boldsymbol{\xi}(n)$, $\hat{\sigma}_{h+1} = \kappa \|\boldsymbol{\xi}(n) - \hat{\boldsymbol{\mu}}_r(n)\|$, $\hat{\mathbf{w}}_{h+1}^T = \mathbf{e}(n)$. Due to the localization property of the $(h+1)$th new hidden neuron, the error will be brought back into the tunable area and E_a decrease since $\epsilon_H(h+1) < \epsilon_H(h)$.

3. Simulation Results

The proposed identification scheme is applied in this section to identify several nonlinear systems. The first one is a discrete nonlinear single-input single-output (SISO) system, and the second is carefully selected to represent a time-varying nonlinear multi-input multi-output (MIMO) missile dynamics. In the simulation, a comparison study between fully tuned RBF networks and a RBF network with only tuning the weights is presented. The network candidates used in this study are,

- **Type-1**: Growing RBF network (GRBFN) with only tuning the weights.
- **Type-2**: GRBF network with all the parameters being adaptable.
- **Type-3**: Growing and Pruning RBF Network (GAP RBFN) with all the parameters adaptable.

The GAP RBFN is the incorporation of GRBFN with a pruning strategy. For a detailed description of the pruning strategy, refers to Chapter 4.

3.1. Example 1: Identification of SISO Nonlinear System

The first example is selected to represent a SISO nonlinear dynamic system. The system to be identified is described by the first order difference equation [43]:

$$x(n) = \frac{29}{40}\sin(\frac{16u(n-1)+8x(n-1)}{3+4u(n-1)^2+4x(n-1)^2}) + \frac{2}{10}u(n-1)+\frac{2}{10}x(n-1) \quad (2.37)$$

where $x(0) = 0$ and n is the time index, and a random signal uniformly distributed in the interval [-1,1] is used for $u(n)$ in the system. A GRBF network is used to identify Eq.(2.36), where the network input $\boldsymbol{\xi} = [x(n-1), u(n-1)]^T$. To guarantee the stability of the overall scheme, the matrix $\mathbf{A}$ is selected as -0.1. In this case, we select the GRBF network's 3 criteria gates as: $E_1 = 0.01, E_2 = 0.01, E_3 = \max(\varepsilon_{max} \times \gamma^n, \varepsilon_{min})$, $\varepsilon_{max} = 1.15, \varepsilon_{min} = 0.4, \gamma = 0.999$, We use $\delta = 0.001$ as the pruning threshold, and the size of the two sliding windows (N_w, S_w) are selected as 48 (S_w is used for pruning the hidden neurons). The network starts with no hidden neuron.

The performance of three typies of RBFN are compared, that is, Type-1, Type-2 and Type-3 respectively. The evolution of the hidden neurons is shown in Fig.2.2. From Fig.2.2, we can see clearly how the hidden neurons are added and pruned according to the criteria. The selection of E_1, E_2 and E_3 directly affects the result of the simulation. If these thresholds are larger than necessary, then the hidden neurons can not be added even there is a requirement, and hence

Nonlinear System Identification Using Lyapunov-Based Fully Tuned RBFN 39

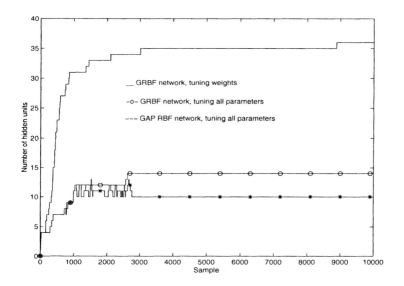

Figure 2.2. Example 1: evolution of hidden neurons

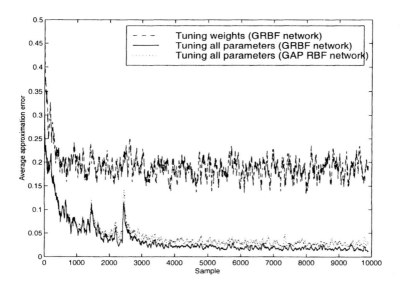

Figure 2.3. Example 1: evolution of average error

the network can not approximate the system dynamics accurately; On the other hand, if the thresholds are set to very small numbers, then too many hidden neuron will be recruited which will lead to a slow convergence speed and even overfitting. According to experience, generally E_1 and E_2 are set to around 1% of the output. The results show that with tuning all the parameters, only 14 and 10 hidden units are used in the steady state (14 for the GRBF network, 10 for the GAP RBF network), while with only tuning the weights, this number is 36.

For comparison purposes, we use the error criteria defined as:

$$E_{ave}(n) = \sqrt{\frac{1}{N_w} \sum_{j=n-N_w+1}^{n} \|x(j) - \hat{x}(j)\|} \qquad (2.38)$$

Fig.2.3 shows the evolution of the error. With tuning weights, the average error is quite large. However, tuning all the parameters can remedy this shortcoming, and from the figure we can see the average error achieved is comparatively less. In addition, although with the GAP strategy, only 10 hidden neurons are recruited, 4 less than with the growing strategy, from the figure we can see only minor difference between average errors of the two approaches.

A quantitative comparison of the number of hidden units and the average identification error is given in Table 2.1.

Table 2.1. Example 1: Comparison of the performance

algorithm	structure	tuning law	hidden unit	E_{ave}
Type-1	GRBF	weights	35	0.20
Type-2	GRBF	all	14	0.04
Type-3	GAP RBF	all	10	0.05

The above results clearly demonstrate the theoretical results that with tuning all the parameters using the derived stable adjusting laws, the performance is much better with far fewer hidden neurons.

3.2. Example 2: Identification of Nonlinear Time-Varying Missile System

In practice, most real systems exhibit complex nonlinear characteristics and can not be treated satisfactorily using linear identification theory, especially the high performance fighter aircraft or a complicated missile dynamics undergoing complex maneuvers. In this section, we use the derived Lyapunov-based tuning rules for fully tuned RBFN to identify a single-input multi-output (SIMO) missile dynamics.

The dynamics of a missile system, similar to the one described in [88], is given in Eq.(2.39). However, to make the system time-varying, a coefficient ζ

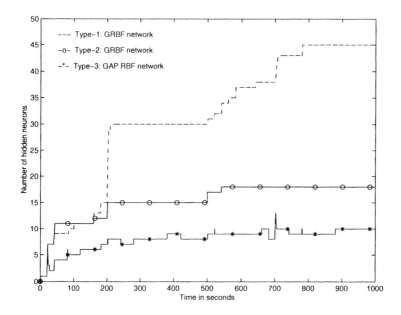

Figure 2.4. Example 2: evolution of hidden neurons

is incorporated.

$$\begin{aligned}\dot{x}_1 &= x_2 - 0.1\zeta\cos\zeta x_1(5x_1 - 4x_1^3 + x_1^5) - 0.5\cos x_1 u\\ \dot{x}_2 &= -65x_1 + 50x_1^3 - 15x_1^5 - x_2 - 100u\end{aligned} \quad (2.39)$$

In this study, the sampling time T_s is selected as 0.1 second, and the input signal is composed of 50 randomly valued step signals, each with a period of 20 seconds and magnitude within $[-0.3, 0.3]$ so that the slide-slip angle $x_1 \leq \pi/2 rad$, x_2 is $\dot{\beta}$ and ζ changes from 1 to 0.8 after 500 seconds, which indicates the missile's dynamics is suddenly changed in 500th second. The performance of RBF networks, Type-1, Type-2 and Type-3 are compared according to the network size and approximation error. To guarantee the stability of the overall scheme, the matrix $\mathbf{A}$ has to be selected *a priori*. The design of $\mathbf{A}$ is related to the system dynamics, in real applications, $\mathbf{A}$ can also be selected such that its eigenvalues result in good overall system response. In this case it is chosen as a $diag(-3, -3)$. As indicated in section 2.2, to make a compromise between approximation accuracy and network size, the dead zone e_0 is set to 0.01, and Z is selected as 1.5. In addition, the GRBF network's 3 criteria gates is selected as: $E_1 = 0.01, E_2 = 0.02, E_3 = \max(\varepsilon_{max} \times \gamma^i, \varepsilon_{min}), \varepsilon_{max} = 1.15, \varepsilon_{min} =$

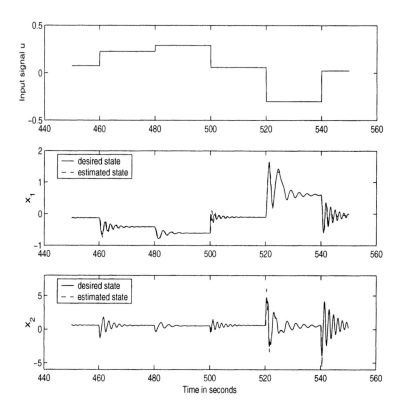

Figure 2.5. Example 2: performance of identification (450-550sec)

$0.4, \gamma = 0.997$, and i is the sample index. $\delta = 0.005$ is used as the pruning threshold, and the size of the two sliding windows(N_w, S_w) as 28.

The evolution of the hidden neurons is shown in Fig.2.4, from which it can be seen clearly how the hidden neurons are added and pruned. The results show that with tuning all the parameters, after reaching steady state only 18 and 10 hidden units are used (18 for the Type-2, 10 for the Type-3), while using GRBF network Type-1, this number is 45. The most prominent phenomenon observed is the change of the hidden neurons together with ζ. When the system changes with ζ, the parameters of neurons acquired before the change may not be correct. With a fully tuned GRBF network (Type-2 and Type-3), those parameters will be updated to adapt the changed dynamics, while with only tuning the weights (Type-1), new hidden neurons have to be added continuously.

To clearly show the proposed fully tuned GRBF network's behavior (Type-2) at $t = 500$ second, in Fig.2.5, the estimated states of x_1 and x_2 compared to the

Nonlinear System Identification Using Lyapunov-Based Fully Tuned RBFN 43

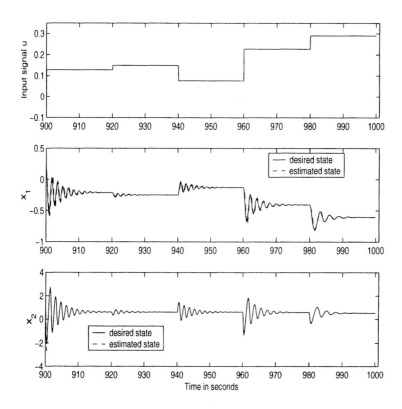

Figure 2.6. Example 2: performance of identification (900-1000sec)

Table 2.2. Comparison of the performance

algorithm	structure	tuning law	hidden unit	E_{ave}
Type-1	GRBF	weights	45	0.286
Type-2	GRBF	all	18	0.058
Type-3	GAP RBF	all	10	0.062

outputs of the missile dynamics from $450sec$ to $550sec$ is depicted. From Fig.2.5 it can be seen that in the period of $450sec$ to $500sec$, the identification is good, while after $500sec$, since the missile dynamics suddenly changes, the GRBF network has to perform on-line learning, therefore the result is not so good. However, after some time, the varied dynamics is also successfully learned,

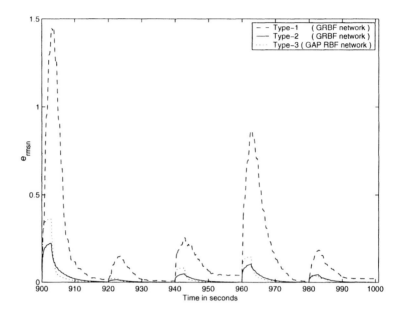

Figure 2.7. Evolution of RMS error e_{rmsn}

which is reflected in Fig.2.6, where the estimated states *vs.* the real outputs of the system from 900sec to 1000sec is shown and they are very close.

The accuracy of the identification using the different strategies is also assessed. To compare the approximation error, the error criteria e_{rmsn} defined in the previous section is used. To increase the clarity, result from $900sec$ to $1000sec$ is presented in Fig.2.7 as before. It is obvious that the performance of Type-1 is not good, but with Type-2 or Type-3, the results are much better. In addition, with GAP strategy, although only 10 hidden neurons are recruited using Type-3, from the figure it is difficult to see any visible difference in average error between Type-2 and Type-3.

A quantitative comparison of all the three methods is given in Table 2.2. The data are collected based on all the 10,000 samples (1000 seconds). The error used in this table is defined as $E_{ave} = \sum_{i=1}^{10000} e_{rmsn}(i)/10000$. From the table, it may be seen that with tuning all the parameters, Type-2/Type-3 achieves better performance than Type-1 with less hidden neurons.

4. Summary

In this chapter, a stable on-line identification scheme using a fully tuned GRBF network is developed. Parameter tuning laws are derived using Lyapunov synthesis approach, and a dead zone is designed to guarantee the convergence of the identification error. Simulation results demonstrate that with tuning all the parameters, the performance is much better. The results also indicate that a GAP RBF network can implement a more compact network structure than GRBF network with a similar accuracy. However, this does not come with a strict mathematical proof on the stability and convergence.

In addition to the proposed Lyapunov based tuning law, nowadays, there are other algorithms for training the fully tuned RBFN. Among them, the minimal resource allocating algorithm (MRAN) with extended Kalman Filter (EKF) tuning rule has been demonstrated to be powerful in the field from pattern recognition to system identification [118]. Therefore in the next chapter, we will further investigate the performance of MRAN algorithm in real-time identification of nonlinear system.

Chapter 3

REAL-TIME IDENTIFICATION OF NONLINEAR SYSTEMS USING MRAN/EMRAN ALGORITHM

1. Introduction

Minimal resource allocating network (MRAN) is a recent algorithm for implementing fully tuned RBF network. Unlike the derived parameter tuning rules in Chapter 3, in MRAN algorithm, an extended Kalman filter (EKF) is utilized to update all the parameters of the RBFN. Although lacking a strict mathematical proof, MRAN was shown to be more effective than other algorithms (like RAN and RANEKF) in function approximation and pattern classification [60].

Recently another sequential learning algorithm for fully tuned RBFN has been proposed by Junge and Unbehauen [43]. Their algorithm incorporates the idea of on-line structural adaptation to add new hidden neurons and uses an error sensitive clustering algorithm to adapt the center and width of the hidden neurons. The algorithm known as On-line Structural Adaptive Hybrid Learning (ONSAHL) is shown in [43] to produce compact networks for nonlinear dynamic system identification problems.

In this chapter, first the performance evaluation of MRAN and ONSAHL algorithm for the same nonlinear identification benchmark problems from [43] and [42] is carried out. This study is intended to compare the complexity of the resulting networks and the accuracy of approximation using MRAN and ONSAHL in the field of nonlinear system identification. Then the performance of the MRAN algorithm in the real-time identification is analyzed and the algorithm is extended to a so called Extended MRAN (EMRAN) algorithm. Comparison studies show that the computation speed of EMRAN is much faster and hence is suitable for real-time implementation.

2. Introduction of MRAN Algorithm

MRAN algorithm, which is proposed by Lu, Sundararajan and Saratchandran [60], combines a pruning strategy with the growth criteria of the RAN algorithm to realize a minimal network structure. This algorithm is an improvement to the RAN of Platt [94] and the RANEKF algorithm of Kadirkamanathan [44]. In this chapter, the MRAN algorithm for training the RBF networks under a discrete framework is reviewed [60] with applications to nonlinear system identification.

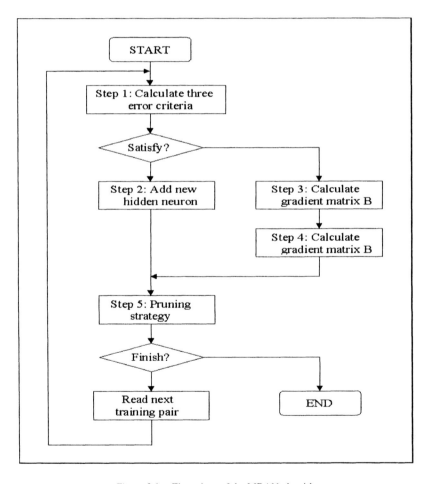

Figure 3.1. Flow chart of the MRAN algorithm

Rewrite the outputs of the Gaussian RBF neural network as,

$$\mathbf{f}(\boldsymbol{\xi}) = \underline{\mathbf{w}}_0^T + \sum_{k=1}^{h} \underline{\mathbf{w}}_k^T \phi_k(\boldsymbol{\xi}) \qquad \boldsymbol{\xi} \in R^{n_x} \qquad (3.1)$$

$$\phi_k(\boldsymbol{\xi}) = \exp(-\frac{1}{\sigma_k^2}\|\boldsymbol{\xi} - \boldsymbol{\mu}_k\|^2) \qquad (3.2)$$

For consistence, $\boldsymbol{\xi}$ is the input vector of the network, h indicates the total number of hidden neurons, the coefficient $\underline{\mathbf{w}}_k$ is the connection weights of the kth hidden neuron and $\underline{\mathbf{w}}_0$ is the bias term. $\boldsymbol{\mu}_k$ and σ_k refer to the center and width of the kth hidden neuron. Generally, a wide class of MIMO nonlinear dynamic systems can be represented by the nonlinear discrete model with an input-output description form,

$$\mathbf{y}(i) = \mathbf{g}[\mathbf{y}(i-1), \cdots, \mathbf{y}(i-m), \mathbf{u}(i-1), \cdots, \mathbf{u}(i-n)] \qquad (3.3)$$

where $\mathbf{y}$ is a vector containing n_y system outputs, $\mathbf{u}$ is a vector for n_u system inputs; $\mathbf{g}[\cdots]$ is a nonlinear vector function, representing n_y hypersurfaces of the system, and m and n are the maximum lags of the output and input, respectively. Selecting $[\mathbf{y}(i-1), \cdots, \mathbf{y}(i-m), \mathbf{u}(i-1), \cdots, \mathbf{u}(i-n)]$ as the neuron network's input $\boldsymbol{\xi}$, essentially the problem for the identification of a nonlinear dynamic system can be converted into a nonlinear time series problem with a one step ahead prediction.

The principle of MRAN algorithm can be simply illustrated using Fig.3.1.

In MRAN algorithm, the network starts with no hidden neuron, and as the input-output training sample $\boldsymbol{\xi}(i), \mathbf{y}(i)$ (i is the time index) is received, the network is built up based on certain growth criteria. The following steps describe the basic ideas of the MRAN algorithm.

- **Step 1:** Calculation of the three error criteria
  The first step of the algorithm is to check whether the criteria for recruiting a new hidden unit are met. They are,

$$\|\mathbf{e}_i\| = \|\mathbf{y}_i - \hat{\mathbf{y}}_i(\boldsymbol{\xi}_i)\| > E_1 \qquad (3.4)$$

$$e_{rmsi} = \sqrt{\sum_{j=i-(N_w-1)}^{i} \frac{\|\mathbf{e}_j\|^2}{N_w}} > E_2 \qquad (3.5)$$

$$d_i = \|\boldsymbol{\xi}_i - \boldsymbol{\mu}_{ir}\| > E_3 \qquad (3.6)$$

where $\boldsymbol{\mu}_{ir}$ is the center of the hidden unit that is closest to current input $\boldsymbol{\xi}_i$. E_1, E_2 and E_3 are thresholds to be selected appropriately. Eq.(3.4) decides if the existing nodes are insufficient to obtain a network output that meets the error specification. Eq.(3.5) checks whether the network meet the required sum squared error specification for the past N_w outputs of the network. Eq.(3.6) ensures that the new node to be added is sufficiently far from all the existing nodes.

Only when all these criteria are met, a new hidden node is added. Then go to Step 2 to add a new RBF hidden unit, otherwise go to Step 3 to update all the parameters of the network using EKF.

- **Step 2**: Inclusion of a new RBF hidden unit
  When all the criteria in Step 1 are satisfied, a new hidden unit is added. Each new hidden unit added to the network will have the following parameters associated with it:

$$\underline{\mathbf{w}}_{h+1}^T = \mathbf{e}_i, \quad \boldsymbol{\mu}_{h+1} = \boldsymbol{\xi}_i, \quad \sigma_{h+1} = \kappa \|\boldsymbol{\xi}_i - \boldsymbol{\mu}_{ir}\| \tag{3.7}$$

These parameters are set to remove the error. The overlap of the responses of the hidden units in the input space is determined by κ, the overlap factor. After adding the new hidden neuron, go to Step 5 to perform a pruning strategy.

- **Step 3**: Calculating the gradient matrix $\mathbf{\Pi}_i$
  If the three criteria for adding a new hidden unit can not be satisfied, then an adaptation of the network parameters should be done. $\mathbf{\Pi}_i = \nabla_\mathbf{w} \mathbf{f}(\boldsymbol{\xi}_i)$ is the gradient vector of the function $\mathbf{f}(\boldsymbol{\xi}_i)$ with respect to the parameter vector χ evaluated at χ_{i-1}, which will be used in the next step.

$$\begin{aligned}\mathbf{\Pi}_i = & [\mathbf{I}, \phi_1(\boldsymbol{\xi}_i)\mathbf{I}, \phi_1(\boldsymbol{\xi}_i)(2\underline{\mathbf{w}}_1^T/\sigma_1^2)(\boldsymbol{\xi}_i - \boldsymbol{\mu}_1)^T, \\ & \phi_1(\boldsymbol{\xi}_i)(2\underline{\mathbf{w}}_1^T/\sigma_1^3)\|\boldsymbol{\xi}_i - \boldsymbol{\mu}_1\|^2, \ldots, \\ & \phi_1(\boldsymbol{\xi}_h)\mathbf{I}, \phi_h(\boldsymbol{x}i_i)(2\underline{\mathbf{w}}_h^T/\sigma_h^2)(\boldsymbol{\xi}_i - \boldsymbol{\mu}_h)^T, \\ & \phi_h(\boldsymbol{\xi}_h)(2\underline{\mathbf{w}}_h^T/\sigma_h^3)\|\boldsymbol{\xi}_h - \boldsymbol{\mu}_h\|^2 \,]^T\end{aligned} \tag{3.8}$$

After this preparation, the vector χ can be updated, and then go to Step 4.

- **Step 4**: Updating the parameters using EKF
  In this step, the network parameters $\chi = [\underline{\mathbf{w}}_0, \underline{\mathbf{w}}_1, \boldsymbol{\mu}_1^T, \sigma_1, \ldots, \underline{\mathbf{w}}_h, \boldsymbol{\mu}_h^T, \sigma_h]^T$ are adapted using the EKF as follows,

$$\chi_i = \chi_{i-1} + \mathbf{K}_i \mathbf{e}_i \tag{3.9}$$

where $\mathbf{K}_i$ is the Kalman gain vector given by,

$$\mathbf{K}_i = \mathbf{P}_{i-1}\mathbf{\Pi}_i[\mathbf{R}_i + \mathbf{\Pi}_i^T \mathbf{P}_{i-1}\mathbf{\Pi}_i]^{-1} \tag{3.10}$$

$\mathbf{R}_i$ is the variance of the measurement noise. $\mathbf{P}_i$ is the error covariance matrix which is updated by,

$$\mathbf{P}_i = [\mathbf{I}_{z \times z} - \mathbf{K}_i \mathbf{\Pi}_i^T]\mathbf{P}_{i-1} + q\mathbf{I}_{z \times z} \tag{3.11}$$

q is a scalar that determines the allowed random step in the direction of the gradient vector. If the number of parameters to be adjusted is z, then $\mathbf{P}_i$ is a $z \times z$ positive definite symmetric matrix. When a new hidden neuron is allocated, the dimensionality of $\mathbf{P}_i$ increases to,

$$\mathbf{P}_i = \begin{pmatrix} \mathbf{P}_{i-1} & 0 \\ 0 & p_0 \mathbf{I}_{z_1 \times z_1} \end{pmatrix} \quad (3.12)$$

where the new rows and columns are initialized by p_0. p_0 is an estimate of the uncertainty when the initial values assigned to the parameters. The dimension z_1 of the identity matrix $\mathbf{I}$ is equal to the number of new parameters introduced by the new hidden neuron. Then go to Step 5 for a pruning strategy.

- **Step 5**: Pruning strategy
The last step of the algorithm is to prune those hidden neurons that contribute little to the network's output for S_w consecutive observations. Let matrix $\mathbf{O}$ denotes the outputs of the hidden layer and $\mathbf{W}$ denotes the weight matrix, consider the output $o_{nj}(j = 1 \cdots n_y)$ of the nth hidden neuron,

$$o_{nj} = w_{nj} \exp(-\frac{1}{\sigma_n^2} \|\xi - \mu_n\|^2), \quad (3.13)$$
$$n = 1 \cdots h, \quad j = 1 \cdots n_y$$

If w_{nj} or σ_n in the above equation is small, o_{nj} might become small. Also, if $\|\xi - \mu_n\|$ is large, the output will also be small. This would mean that the input is sufficiently far away from the center of this hidden neuron. To reduce the inconsistency caused by using the absolute value of the output, this value is normalized to that of the highest output,

$$r_{nj} = \frac{o_{nj}}{\max\{o_{1j}, o_{2j}, \cdots, o_{hj}\}}, \quad (3.14)$$
$$n = 1 \cdots h, \quad j = 1 \cdots n_y$$

The normalized output of each neuron r_{nj} is then observed for S_w consecutive inputs. A neuron is pruned, if its output $r_{nj}(j = 1 \cdots n_y)$ falls below a threshold value(δ) for the S_w consecutive inputs. Then the dimensionality of all the related matrices will be adjusted to suit the reduced network.

The sequential learning algorithm for MRAN, can be summarized as follows.

- Obtain an input and calculate the network output (Eq.(1.8)) and the corresponding errors (Eq.(3.4) to Eq.(3.6)).
- Create a new RBF center (Eq.(3.7)) if all the three inequality Eq.(3.4) to Eq.(3.6) hold.

- If criteria for adding neurons are not met, adjust the weights and widths of the existing RBF network using EKF (Eq.(3.8) to Eq.(3.12)).

- Last a pruning strategy is adopted to prune those hidden neurons that contribute little to the output for a certain number of consecutive samples (Eq. (3.14)).

A number of successful applications of MRAN in different areas such as pattern classification, function approximation and time series prediction have been reported in [59][61].

3. Performance of the MRAN Algorithm

In this section, we compare the performance of MRAN with the ONSAHL algorithm for two benchmark problems on nonlinear dynamic systems identification used in [43]. For convenience, these two benchmark problems are referred to as BM-1 and BM-2. A brief recount of ONSAHL algorithm is given below.

3.1. The ONSAHL Algorithm

The ONSAHL algorithm has been proposed especially for identifying time varying nonlinear dynamic systems [43]. The ONSAHL algorithm uses an extended RBF network, which is referred to as a Direct Linear Feedthrough RBF (DLF-RBF). DLF-RBF is composed of a nonlinear RBF sub-network and a linear DLF sub-network part, which are connected in parallel, performing a mapping from input layer directly to the output layer. The ONSAHL algorithm uses the same growth criteria as in Platt's RAN but differs in the way the centers and widths are adjusted. Unlike RAN where the centers of all hidden neurons are updated to fit the training data (in the LMS sense), in the ONSAHL only the center and width of the neuron which is nearest to the input are updated in the first step. Then all the weights connected to the output layer are updated as in the case of RAN using the RLS method. ONSAHL algorithm has been tested on two single input single output (SISO) nonlinear time-invariant and time varying dynamic system identification problems, i.e. BM-1 and BM-2. The following steps describe the basic ideas of the ONSAHL algorithm.

- **Step 1:** Start the algorithm initially from a network without RBF neurons, and gradually include a new neuron to the model. If the condition for adding new hidden neurons can not be met, then select a hidden neuron which has the minimum distance to the input vector, then go to Step 1.2 and Step 1.3 to adapt the parameters of that hidden neuron. Otherwise go to Step 1.1 to add a new hidden neuron.

- **Step 1.1:** Include a new RBF neuron, then go to Step 2.

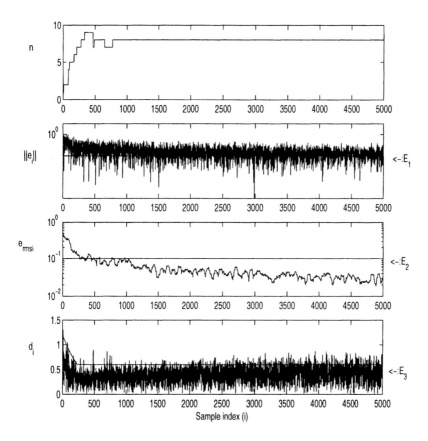

Figure 3.2. BM-1: performance of MRAN algorithm

- **Step 1.2:** Adapt the center vector of the hidden neuron.

- **Step 1.3:** Adapt the width vector of the hidden neuron.

- **Step 2:** Adapt the output layer weights of the network.

- **Step 3:** Stop the calculation if the desired minimal error value is reached, otherwise go to Step 1 again.

For a detailed description of the ONSAHL algorithm, see [43].

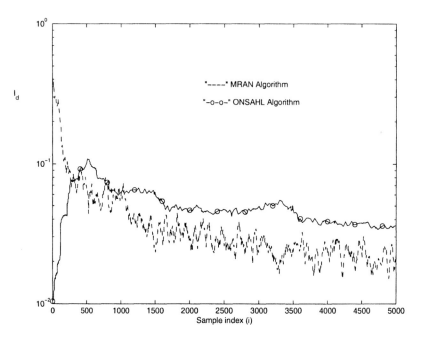

Figure 3.3. BM-1: evolution of error(I_d) (MRAN vs. ONSAHL)

3.2. BM-1: Nonlinear SISO Time-Invariant System

The nonlinear SISO time-invariant system (BM-1) to be identified is described by the following first order difference equation,

$$x(i) = \frac{29}{40}\sin(\frac{16u(i-1) + 8x(i-1)}{(3 + 4u(i-1)^2 + 4x(i-1)^2})$$
$$+ \frac{2}{10}u(i-1) + \frac{2}{10}x(i-1) \qquad (3.15)$$

A random signal uniformly distributed in the interval $[-1, 1]$ is used for $u(i)$ in the system. In order to analyze the factual approximation to the system's input-output mapping, a so-called deterministic on-line error index $I_d(i)$ is defined,

$$I_d(i) = \frac{1}{N_w}\sum_{p=0}^{N_w-1}|x(i-p) - \hat{x}(i-p)| \qquad (3.16)$$

N_w is the past N_w times outputs of the network, and $\hat{x}$ is the output of identification. The performance of the MRAN algorithm and the ONSAHL algorithm is then compared according to the error I_d.

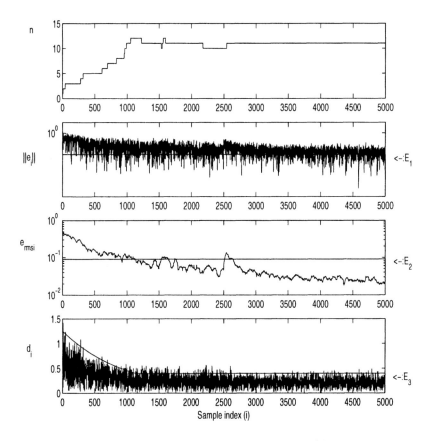

Figure 3.4. BM-2: performance of MRAN algorithm

In this case, the MRAN's 3 criteria gates are selected as: $E_1 = 0.01, E_2 = 0.1, E_3 = \max(\varepsilon_{max} \times \gamma^i, \varepsilon_{min}), \varepsilon_{max} = 1.2, \varepsilon_{min} = 0.6, \gamma = 0.997$, we use $\delta = 0.0001$ as the pruning threshold, and the size of the two sliding windows (N_w, S_w) as 48.

The identification results using MRAN and ONSAHL are given in Fig.3.2 and Fig.3.3. Fig.3.2 presents the hidden neuron evolution history along with the time history of the three error functions. From Fig.3.2 one can see clearly how the hidden neurons are added and pruned according to the three criteria. In this case, MRAN takes 8 hidden units to identify the system, while from [43], the ONSAHL takes 23 hidden neurons for the same problem. Fig.3.3 shows the values of the error index I_d time history. It can bee seen from Fig.3.3 that

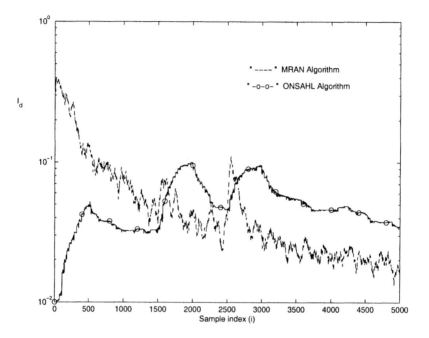

Figure 3.5. BM-2: evolution of error(I_d) (MRAN vs. ONSAHL)

MRAN's error is coming down quickly and in the steady state it is lower than that of ONSAHL.

3.3. BM-2: Nonlinear SISO Time-Varying System

The second example BM-2 is selected to identify a nonlinear SISO time-varying discrete system given as:

$$\begin{aligned}x(i) &= \frac{29\beta(i)}{40}\sin(\frac{16u(i-1)+8x(i-1)}{\beta(i)(3+4u(i-1)^2+4x(i-1)^2})\\ &+\frac{2}{10}u(i-1)+\frac{2}{10}x(i-1)\end{aligned} \quad (3.17)$$

$\beta(i)$ is a time-varying parameter given in Table 3.1. The system input signal $u(i)$ used in this example is a random signal with uniform distribution in the interval $[-1, 1]$.

Fig.3.4 and Fig.3.5 present the identification results for this benchmark problem. From Fig.3.4 the most important thing one can observe is the change of the hidden neurons together with β. Briefly speaking, when the system dynamics changes because of the variation in β, the network has to add new neurons to

Table 3.1. Evolution of β (BM-2)

index(i)	0 to 1500	1501 to 2500	2501 to 5000
$\beta(i)$	1.0	0.9	0.8

Table 3.2. Comparison of identification results of the MRAN and ONSAHL

	BM-1		BM-2	
Performance	I_{dav}	Hidden units	I_{dav}	Hidden units
ONSAHL	0.0437	23	0.0586	25
MRAN	0.0261	8	0.0326	11

adapt for the changes. When the change of β neither affect the dynamics nor changes the complexity of the system after some time, the old neuron will be pruned. Compared to the results achieved in [43], MRAN uses only 11 hidden neurons while the ONSAHL uses 25 hidden neurons, and from Fig.3.5 we can see that the MRAN algorithm also achieves better approximation result, that is, I_d is smaller. It can be seen that the error in Fig.3.5 increases and decreases together with β and the number of hidden neurons. The approximation error suddenly increases when the system dynamics changes at step number 1,500 and 2,500. At the same time the network adds new hidden neurons to decrease this error. After some time when the changed system dynamics is learnt by the network, the old neuron is pruned and the approximation error decreases again. Further, for this example, MRAN starts with no hidden neuron whereas ONSAHL starts with a trained network from BM-1 (23 hidden neurons).

The identification results for both MRAN and ONSAHL algorithms are compared in Table 3.2. In Table 3.2, I_{dav} is defined as the average value of the I_d calculated from 1,500th sample to 5,000th sample. It is selected in this way because from the figure one can see that after 1,500 samples, the identification results converges to a comparatively smooth one.

Based on these two problems, it can be concluded that MRAN is able to perform better with a smaller network structure. However, for practical use, MRAN's on-line implementation issues have to be analyzed. Any improvements to MRAN minimizing its computational effort is always welcome, such a modification is described in the following sections.

4. Real-Time Implementation of MRAN and the Extended MRAN Algorithm

4.1. Analysis of MRAN for Real-Time Implementation

For practical purposes, it should be possible to run any identification algorithm in real-time and MRAN is also no exception. Since MRAN uses sequential learning scheme, for real-time implementation, the learning time t_c for one set of input and output data must be less than the sampling time τ selected for identification. In this section, we estimate MRAN algorithm's learning time t_c for one cycle based on its detailed breakdown into a number of computational steps done in one cycle. All the time estimates presented here are based on running MRAN on a Pentium 120MHz computer under the environment of $VC^{++}5.0$ in Win95 with no other programs running.

Looking at MRAN equations, it is clear that t_c mainly consists of 5 parts, corresponding to the Steps 1 to 5 described before. The computation time for step i is referred to as t_{c_i} and the cycle time t_c is given by

$$t_c = \sum_{i=1}^{5} t_{c_i} \qquad (3.18)$$

For this timing study, the benchmark problem BM-2 has been selected as the first candidate. Since the time for training one sample of data is determined by factors such as RBF network's structure (number of inputs, number of outputs), and the number of the hidden neurons, for finding out the t_c the network size was varied from 5 to 35 hidden neurons in an increment of 5 without really worrying about the approximation accuracy. *i.e.* The following question was posed : if the MRAN network had 10 hidden neurons for the problem of BM-2 what are the constituent times t_{c_i}s without really worrying about whether the 10 hidden neuron network produced a good approximation. This approach makes the timing analysis easier as MRAN really produces a network with varying number of hidden neurons and to calculate times based on this will be difficult. For the benchmark problem of BM-2, the breakdown for all the times in ms for the 5 steps is given in Table 3.3.

In Table 3.3, the last row gives the total time t_c. Although it is not correct to use 'total' time as the training time for each sample of data (For example, if a new training pair (ξ_i, y_i) satisfie the condition for recruiting a new hidden unit, then after the hidden neuron is added, it directly goes to pruning strategy of Step 5, therefore the t_c is only the sum of t_{c1}, t_{c2} and t_{c5}.), this calculation considers the 'worst' case scenario for calculating 't_c'.

Even though the BM-2 problem gives some idea about the MRAN computation cycle times, the BM-2 problem is still a SISO nonlinear system though time-varying. A realistic assessment can only be made if the problem selected is of a reasonably larger size like a MIMO problem. In this context, the 2 in-

Table 3.3. Computation cycle time $t_c(ms)$ for MRAN (BM-2 problem)

Hidden units	5	10	15	20	25	30	35
t_{1c}	0.82	1.37	2.07	2.38	3.00	3.65	4.55
t_{2c}	0.003	0.004	0.004	0.004	0.005	0.005	0.005
t_{3c}	1.10	2.70	3.32	4.80	5.97	7.32	8.97
t_{4c}	6.14	30.97	105.5	269.7	360.9	533.2	936.0
t_{5c}	0.09	0.17	0.26	0.33	0.44	0.53	0.67
t_c	8.15	35.21	111.2	277.2	370.3	544.7	950.2

puts 2 outputs nonlinear system identification problem (example 2 of [42]) is selected here as the BM-3 problem.

4.2. BM-3: Nonlinear MIMO Time-Invariant System

The MIMO nonlinear dynamic system is given by:

$$x_1(i) = \frac{15u_1(i-1)x_2(i-2)}{2+50u_1(i-1)^2} + \frac{1}{2}u_1(i-1) - \frac{1}{4}x_2(i-2) + \frac{1}{10}$$

$$x_2(i) = \frac{\sin[\pi u_2(i-1)x_1(i-2)] + 2u_2(i-1)}{3} \qquad (3.19)$$

The two random input signals $u_1(i), u_2(i)$ uncorrelated with each other and uniformly distributed in the interval $[-1, 1]$, are used to generate the on-line training set together with the output. Once the plant is identified by the MRAN network, for testing the accuracy of the identified model, the signals shown in Fig.3.6 are used as inputs. These signals have both frequency modulation and amplitude modulation so that they can test the RBFN's generalization ability and adaptability to the data's oscillation.

Table 3.4. Computation cycle time $t_c(ms)$ for MRAN (BM-3 problem)

Hidden units	5	10	15	20	25	30	35
t_{1c}	1.51	2.63	3.87	5.13	7.11	8.07	9.37
t_{2c}	0.005	0.005	0.008	0.008	0.009	0.009	0.01
t_{3c}	1.60	3.67	5.87	8.13	9.70	11.98	14.28
t_{4c}	22.2	154.3	504.1	1202	2080	3548	6195
t_{5c}	0.10	0.18	0.26	0.36	0.46	0.55	0.68
t_c	25.42	160.8	514.1	1215	2097	3568	6219

Table 3.4 gives the computational cycle times for the 5 steps with varying hidden neurons. It can be seen directly that because the number of inputs

60   FULLY TUNED RBF NEURAL NETWORK FOR FLIGHT CONTROL

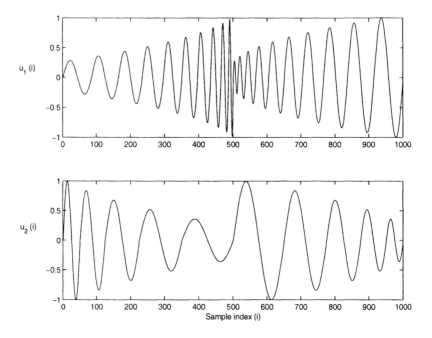

Figure 3.6.   BM-3: test input signals

and outputs for the system increases, the cycle times for BM-3 is considerably higher specially for the case of higher number of hidden neurons. From Table 3.4 it is also evident that Step 4 is the real bottleneck and consumes a large of chunk of computational overhead.

From both Table 3.3 and Table 3.4, it can be seen that with both the increase of the network inputs and outputs and also the number of hidden neurons, time consumed in Step 4 (tuning parameters using EKF) is large. For example, in BM-3 problem, when the hidden neuron reaches 30, from Table 3.4 we can see that Step 4 consumes more than 99% of the total training time, and to satisfy $t_c < \tau$, τ must be larger than 3.6 second, which is unacceptable for many real systems.

To find out reasons for why the time consumed in Step 4 is large with the increase in the hidden neurons , one can take a close look at the computations involved in Step 4, *i.e.* Eq.(3.9) to Eq.(3.11). Because multiplication consumes more time than addition and subtraction, it is worth looking at the number of multiplications involved in Step 4. Note that when matrix $\mathbf{U}_{c \times d}$ multiplies matrix $\mathbf{V}_{d \times h}$, the total number of separate multiplication are $c \times d \times h$. The matrices involved in Step 4 (Eq.(3.9) to Eq.(3.11)) and their sizes are,

$\mathbf{K}_i : (S, n_y)$; $\mathbf{\Pi}_i : (S, n_y)$; $\mathbf{P}_i : (S, S)$; $\mathbf{w}_i : (S, 1)$; $\mathbf{e}_i : (n_y, 1)$; $\mathbf{R}_i : (n_y, n_y)$. Where S is defined as $S = h \times (n_x + n_y + 1) + n_y$.

With the sizes of the above matrices known, the total number of multiplications needed is,

- to calculate $\mathbf{K}_i \mathbf{\Pi}_i^T \mathbf{P}_{i-1}$: $S^3 + S^2 n_y$

- to calculate Eq.(3.10): $S^2 n_y + S n_y^2 + 4/3 n_y^3 + 3/2 n_y^2 + 19/6 n_y - 3$ (using the LU factorization based general matrix inverse algorithm [49]).

- to calculate $(\mathbf{K}_i \mathbf{e}_i)$: $S n_y$

The total number of multiplications in Step 4 is,

$$Sum(h, n_x, n_y) = S^3 + 2S^2 n_y + S(n_y^2 + n_y) \\ + 4/3 n_y^3 + 3/2 n_y^2 + 19/6 n_y - 3 \quad (3.20)$$

From Eq.(3.20), the total time for Step 4 is a third-order polynomial and is a function of h, n_x, n_y. Hence if the number of neurons, or inputs and outputs increase the computational time for Step 4 will increase enormously.

If MRAN has to be modified for real-time implementation, this bottleneck in Step 4 has to be overcome. Such a modification to MRAN is discussed in the next section.

4.3. Extended MRAN (EMRAN) Algorithm

We see from Eq.(3.20) that the weakness in MRAN which increases its computational load is that all the parameters of the network, including all the hidden neuron's centers, widths and weights have to be updated in every step, so the size of the matrices to be updated become large as the number of hidden neurons increase.

To overcome this bottleneck of MRAN for real-time implementation, a new algorithm called EMRAN, which is an improved version of the MRAN algorithm is proposed in this book. For this purpose a 'winner neuron' strategy is incorporated similar to the one described in ONSAHL algorithm. The key idea of the EMRAN algorithm is that in every step, only those parameters that are related to the selected winner neuron are updated by the EKF algorithm. The 'winner neuron' is defined as the neuron in the network that is closest (in some norm sense) to the current input data as in [42][43]. EMRAN attempts to reduce the online computation time considerably and to avoid the overflow of the memory, retaining at the same time the good characteristics of MRAN, namely less number of hidden neurons, lower approximation error, etc.

Basically EMRAN has the same form as MRAN and all the equations are the same and hence are not repeated here. Only the changes are highlighted: In Eq.(3.7), μ_{ir} is defined as the hidden neuron that is the nearest one to the current input data $\boldsymbol{\xi}_i$ in the input space. Here this special hidden neuron is referred to as the "winner neuron", and the parameters related to this neuron are denoted as $\boldsymbol{\mu}^*$, δ^* and $\underline{\mathbf{w}}^*$. The criteria for adding and pruning the hidden neurons are all the same, the difference is that if the training sample does not meet the criteria for adding new hidden neuron, only the network parameters χ^* related to the winner neuron are updated using the EKF as follows,

$$\chi_i^* = \chi_{i-1}^* + \mathbf{K}_i^* \mathbf{e}_i \qquad (3.21)$$

and $\chi^* = [\underline{\mathbf{w}}_0, \underline{\mathbf{w}}^*, \boldsymbol{\mu}^{*T}, \sigma^*]^T$.

In this equation, $\mathbf{K}_i^*$ is the Kalman gain matrix with the size of $(2n_y + n_x + 1) \times n_y$,

$$\mathbf{K}_i^* = \mathbf{P}_{i-1}^* \mathbf{\Pi}_i^* [\mathbf{R}_i + \mathbf{\Pi}_i^{*T} \mathbf{P}_{i-1}^* \mathbf{B}_i^*]^{-1} \qquad (3.22)$$

$\mathbf{\Pi}_i^* = \nabla_{\mathbf{w}^*} \mathbf{f}(\boldsymbol{\xi}_i)$ is the gradient vector of the function $\mathbf{f}(\boldsymbol{\xi}_i)$ with respect to the parameter vector χ^* evaluated at χ_{i-1}^*:

$$\mathbf{\Pi}_i^* = [\mathbf{I}, \phi^*(\boldsymbol{\xi}_i)\mathbf{I}, \phi^*(\boldsymbol{\xi}_i)\frac{2\mathbf{w}^{*T}}{\sigma^{*2}}[\boldsymbol{\xi}_i - \boldsymbol{\mu}^*]^T, \phi^*(\boldsymbol{\xi}_i)\frac{2\mathbf{w}^{*T}}{\sigma^{*3}}\|\boldsymbol{\xi}_i - \boldsymbol{\mu}^*\|^2]^T \quad (3.23)$$

$\mathbf{R}_i$ is the variance of the measurement noise. $\mathbf{P}_i^*$ is the error covariance matrix which is updated by:

$$\mathbf{P}_i^* = (\mathbf{I} - \mathbf{K}_i \mathbf{\Pi}_i^{*T})\mathbf{P}_{i-1}^* + q\mathbf{I} \qquad (3.24)$$

and the size of $\mathbf{P}_i^*$ is $(2n_y + n_x + 1) \times (2n_y + n_x + 1)$.

Table 3.5. Computation cycle time $t_c(ms)$ for EMRAN (BM-2)

Hidden units	5	10	15	20	25	30	35
t_{1c}	0.82	1.37	2.07	2.38	3.00	3.65	4.55
t_{2c}	0.003	0.004	0.004	0.004	0.005	0.005	0.005
t_{3c}	0.22	0.27	0.23	0.25	0.24	0.25	0.26
t_{4c}	1.67	1.50	1.51	1.42	1.48	1.57	1.54
t_{5c}	0.09	0.17	0.26	0.33	0.44	0.53	0.67
t_c	2.80	3.31	4.07	4.38	5.17	6.00	7.03

Using EMRAN, the cycle time and its breakup for one sample of data are given in Table 3.5 and Table 3.6 for both problems BM-2 and BM-3, respectively. From the tables, it can be seen that there is a large reduction in t_{c_4} while all other times remain the same. This time reduction in Step 4 results in a major reduction for the cycle time t_c for EMRAN.

To look at the computational overhead reduction clearly, the above data is also displayed in Fig.3.7 in a bar graph form. Logarithmic scale has been used

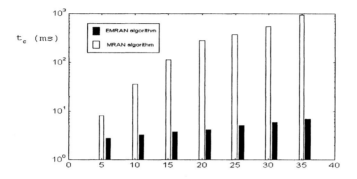

BM-2

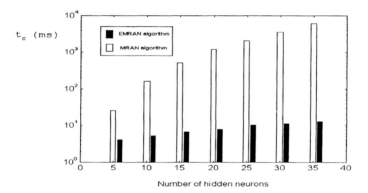

BM-3

Figure 3.7. Comparison of cycle times for MRAN/EMRAN algorithm

for this purpose because the difference between MRAN and EMRAN is large. For BM-2 problem for a typical network with say 30 neurons the cycle time for MRAN is $545ms$ whereas for EMRAN it is $6ms$, a significant reduction. This reduction is more for BM-3 problem as for a network of 30 neurons MRAN takes $3568ms$ whereas EMRAN takes only $11.15ms$, a significant two-order reduction. This behavior is observed for the network with all sizes (number of neurons). It is evident from Fig.3.7 that EMRAN produces a large reduction

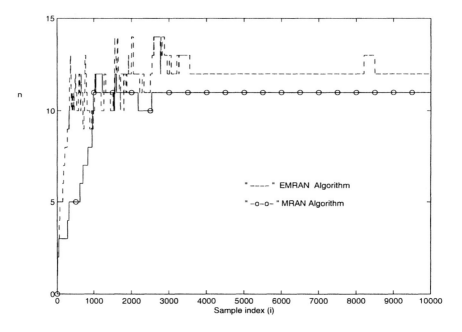

Figure 3.8. BM-2: Evolution of hidden neurons (MRAN vs. EMRAN)

in computation time and also that as the number of hidden neurons increases more than 20 the cycle time remains approximately flat.

Table 3.6. Computation cycle time $t_c(ms)$ for EMRAN (BM-3)

Hidden uints	5	10	15	20	25	30	35
t_{1c}	1.51	2.63	3.87	5.13	7.11	8.07	9.37
t_{2c}	0.005	0.005	0.008	0.008	0.009	0.009	0.01
t_{3c}	0.35	0.38	0.40	0.42	0.40	0.42	0.45
t_{4c}	2.12	2.04	2.03	1.95	2.19	2.11	2.05
t_{5c}	0.10	0.18	0.26	0.36	0.46	0.55	0.68
t_c	4.08	5.23	6.57	7.87	10.17	11.16	12.56

5. Performance Comparison of MRAN vs. EMRAN

The analysis of the last section has looked at cycle times of MRAN and EMRAN without looking at their performances, *i.e.* the identification accuracy

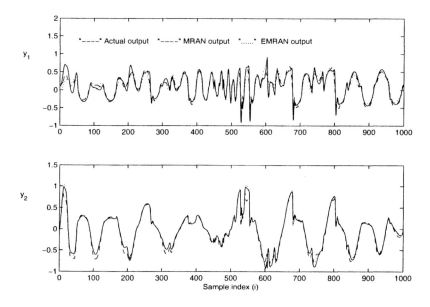

Figure 3.9. BM-3: Test output data (MRAN vs. EMRAN)

of MRAN and EMRAN have not been compared. In this section, the identification accuracy comparison for both the problems BM-2 and BM-3 are carried out using MRAN and EMRAN.

5.1. BM-2: Nonlinear SISO Time-Varying System

The network input and output vector is the same as before. The parameters for the EMRAN are selected as: $E_1 = 0.01, E_2 = 0.09, E_3 = \max(\varepsilon_{max} \times \gamma^i, \varepsilon_{min})$, $\varepsilon_{max} = 1.25, \varepsilon_{min} = 0.4, \gamma = 0.999$, We use $\delta = 0.001$ as the pruning threshold, and the size of the two sliding windows (N_w, S_w) are also 48.

Table 3.7. Comparison of identification results of the MRAN and EMRAN *

Perfor-mance	BM-2				BM-3			
	I_{dav}	Hidden neurons	Overall time (sec)	t_c (ms)	I_{dav}	Hidden neurons	Overall time (sec)	t_c (ms)
MRAN	.0379	11	104.0	111	.0349	30	28574	6219
EMRAN	.0427	12	21.3	3.80	.0392	32	54.8	12.56

* 10,000 samples, t_c is calculated based on 35 hidden units

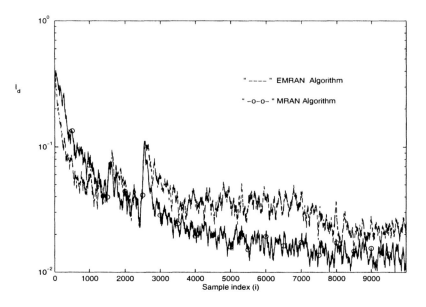

Figure 3.10. BM-2: Evolution of error(I_d)(MRAN vs. EMRAN)

Fig.3.8 to Fig.3.10 give the neuron history and error history for this system. The continuous line indicates the results achieved by the EMRAN algorithm, while the dotted line is that for MRAN. Compared to the neurons for MRAN which is at 11, 12 hidden neurons are used in EMRAN. From Fig.3.10 the approximation errors between the EMARN and MRAN are near and whenever there is a change in the dynamics of the system, although the approximation errors using both of the algorithms increase suddenly, MRAN gives lower errors compared to EMRAN. Table 3.7 presents a comparison of MRAN and EMRAN in terms of the network size, approximation errors and computational times for both problems BM-2 and BM-3. I_{dav} in the table is the average of error index I_d based on 10,000 samples, and for t_c, the network has (maximum) 15 hidden neurons. In the tables, "overall time" represents the total identification time for all the samples and is given in seconds.

5.2. BM-3: MIMO Nonlinear Dynamic System

Fig.3.11 shows the evolution of hidden neurons using MRAN and EMRAN algorithms for BM-3. Although MRAN produces a slightly more compact network than the EMRAN algorithm, from Table 3.7 we can see that the computation time of EMRAN is greatly reduced compared to that of the MRAN algorithm.

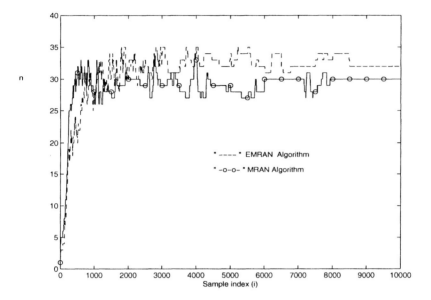

Figure 3.11. BM-3: Evolution of hidden neurons (MRAN vs. EMRAN)

Fig.3.9 gives the results of the identification for both MRAN and EMRAN for problem BM-3. Using the test inputs given in Fig.3.6, the true outputs along with the outputs based on the identified models by MRAN and EMRAN are given in Fig.3.9. We can see from Fig.3.9 that the adding and pruning capability of the EMRAN allows the RBFN to identify the high-dimension nonlinear system on-line. EMRAN produces an output which is closer to MRAN but at a great reduction in computation time and hence can be used for real time identification applications easily.

From the tables it is clear that with a slightly increase of error in EMRAN one gets a great advantage in computational time for EMRAN compared to MRAN.

6. Summary

This chapter has presented a performance analysis of the recently developed Minimal Resource Allocating Network (MRAN) algorithm for on-line identification of nonlinear dynamic systems. Using nonlinear time invariant and time-varying identification benchmark problems, MRAN's performance is compared with the ONSAHL. The results indicate that MRAN realizes networks with far fewer hidden neurons than the ONSAHL algorithm with better approximation accuracy.

Next, the problems in real-time implementation of MRAN has been highlighted using a detailed timing studies and analysis of the basic computations in MRAN. An extension to MRAN referred to as the Extended Minimum Resource Allocating Network (EMRAN) which utilizes a "winner neuron" strategy in MRAN is highlighted. This modification reduces the computation load for MRAN and leads to considerable reduction in the identification time with only a minimal increase in the approximation error. This also indicates the minimum sampling time one can select using EMRAN for identification problems. Using the same benchmark problems as before, the benefits of EMARN show that compared with other learning algorithms EMRAN can "adaptively track" the dynamics of the nonlinear system quickly without loss of accuracy and is ideal for fast on-line identification of nonlinear plants.

Having discussed in detail several identification algorithms using the fully tuned RBFN, in the next chapter, the indirect adaptive control strategies based on the fully tuned RBFN identification schemes are studied.

Chapter 4

INDIRECT ADAPTIVE CONTROL USING FULLY TUNED RBFN

1. Neural Network Based Indirect Adaptive Control

Due to the adaptive characteristics of the learning process, the application of neural networks to nonlinear system identification and control has been developed in a natural way. To cope with the indirect adaptive control problem, two basic approaches have been addressed in the literature. In the first approach, some design problems are learned off-line, measuring the input-output signals and observing the plant behavior in some key situations. The control can then be implemented based on the knowledge acquired: this approach is known as an off-line training/on-line control scheme. In the second approach, an adaptive learning is implemented and the control input is determined on-line as the output of a neural network, which is called on-line learning/on-line control strategy.

Originally, the neural network was used to learn the system dynamics directly. For example, in Narendra's paper [85], a multilayer feed-foreward network (MFN) is used to learn the system dynamics and then an iterative off-line technique is employed to adjust the parameters of the network so as to develop the inverse system model of the plant. However, this method assumed that the form of the nonlinear dynamics of the system is known. Behera *et al.* also explores NN based indirect control for robot manipulator control application [7]. They investigated the application of inversion of a RBF network to nonlinear control problems for which the structure of the nonlinearity is unknown. The RBFN, learning the forward dynamics of the plant, is inverted to represent the inverse dynamics of the system. Based on this inversion, the control structure is proposed and a feedback control law is derived so that the system is Lyapunov stable. Simulation studies show that the performance of the proposed method outperforms Narendra's method and it does not need the prior knowledge of the nonlinear system.

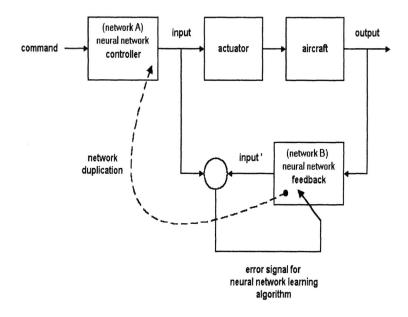

Figure 4.1. Indirect control architecture using neural networks

Instead of learning the forward dynamics of the system, neural networks can also be used directly to learn the inverse dynamics of the system to be controlled. In [105], Sadhukhan and Feteih have presented an exact inverse neural controller with full state feedback that was designed for a linearized F8 aircraft and the architecture is shown to possess stable and near desired responses. For convenience, the proposed architecture in [105] is re-drawn in Fig.4.1 in a simplified version. In their studies, a multi-layer feedforward neural network was trained off-line and used as the controller (network A) as well as the feedback network B. Upon receiving the incoming command signal, the network A outputs the control signal to the aircraft. Because of modeling errors, change in fuel and mass distribution, and sudden changes due to battle damage, the actual aircraft model is bound to differ from the mathematical model used. A neural network feedback identifier (network B) has therefore been included in the control architecture, to learn on-line the inverse dynamics of the aircraft and actuators combined. After the feedback network has updated its parameters based on the error signal, these newly learned characteristic is duplicated into the controller network A and the on-line simulation procedure is repeated for the next sample. The objective of the control strategy is to obtain minimum error such that the output signals of the aircraft follow the pilot's input command signal, that includes the velocity and pitch rate commands in this case.

In this chapter, we investigate Sadhukhan *et al.*'s indirect adaptive control strategy that uses an exact inverse neuro-controller. In this study, the single layer of linear neurons used in the architecture by Sadhukhan and Fetieh is replaced with the RBFN. Moreover, for training the network identifier and controller, the stable identification strategy and the Lyapunov-based tuning rule developed in Chapter 3 as well as the MRAN algorithm discussed in Chapter 4 are applied instead of the gradient descent learning algorithm.

In the first part, simulation results using the off-line training/on-line control scheme is presented. In the off-line training, the training pairs are arranged in several different ways to test the performance. Moreover, in on-line control procedure, the performance of the RBFN controllers are investigated based on different network structures, *i.e.*, the RBFN with a fixed number of hidden neurons, the RBFN with a growing strategy, the RBFN with both a growing and a pruning strategy. In the second part, avoiding the off-line training, the on-line learning/on-line control strategy is explored. The performance of the above schemes for controlling the linearized longitudinal F8 aircraft model are evaluated based on the simulation studies.

2. Neural Network Controller with Off-Line Training and On-Line Control

In this section, the indirect adaptive control strategy with off-line training and on-line control is investigated. An exact inverse neuro-controller with a full state feedback architecture for the linearized longitudinal F8 aircraft model is studied.

2.1. Linearized Longitudinal F8 Aircraft Model

A linearized longitudinal dynamic model of the F8 aircraft given in [105] is used for this study. Since the flight control system is implemented on a digital computer, selecting the sampling time as $\tau = 0.05 sec$, the linearized discrete longitudinal model of F8 aircraft under the following trim condition: $v(0) = 650 ft/s$ (Mach number=0.6), $\alpha(0) = 0.078 rad$, and altitude $20,000 ft$, can be written as,

$$\mathbf{x}(n+1) = \mathbf{F}_1 \mathbf{x}(n) + \mathbf{G}_1 \mathbf{u}(n) \tag{4.1}$$

The state vector $\mathbf{x}$ comprises of the velocity (v in ft/s), angle of attack (α in rad), pitch rate (q in rad/sec) and pitch angle (θ in rad). $\mathbf{u}$ includes δ_{ea} in rad and δ_{ta} in percentage. δ_{ea} denotes the elevator's output and δ_{ta} is the throttle's output. The plant matrices $\mathbf{F}_1$ and $\mathbf{G}_1$ in the model are given by,

$$\mathbf{F}_1 = \begin{bmatrix} 0.9990 & -0.6807 & -0.0570 & -1.6100 \\ 0.0000 & 0.9531 & 0.0483 & 0.0000 \\ 0.0000 & -0.2317 & 0.9700 & 0.0000 \\ 0.0000 & -0.0059 & 0.0493 & 1.0000 \end{bmatrix}$$

$$\mathbf{G}_1 = \begin{bmatrix} -0.0448 & 0.4450 \\ -0.0160 & 0.0000 \\ -0.4282 & 0.0000 \\ -0.0108 & 0.0000 \end{bmatrix}$$

The actuator dynamics $\mathbf{u} = [\delta_{ea}, \delta_{ta}]^T$ is represented by,

$$\begin{bmatrix} \delta_{ea}(n) \\ \delta_{ta}(n) \end{bmatrix} = \mathbf{F}_2 \begin{bmatrix} \delta_{ea}(n-1) \\ \delta_{ta}(n-1) \end{bmatrix} + \mathbf{G}_2 \begin{bmatrix} \delta_{ec}(n-1) \\ \delta_{tc}(n-1) \end{bmatrix} \quad (4.2)$$

where δ_{ec}, δ_{tc} are the input command to the elevator and throttle respectively, and,

$$\mathbf{F}_2 = \begin{bmatrix} 0.6065 & 0 \\ 0 & 0.7788 \end{bmatrix}, \quad \mathbf{G}_2 = \begin{bmatrix} 0.3935 & 0 \\ 0 & 0.2212 \end{bmatrix}$$

The control system is designed to provide decoupled responses to pilot velocity and pitch rate commands. The desired response transfer function as per level 1 handling qualities are given by

$$\frac{v_c}{v_{sel}} = \frac{0.04(s+3.13)}{s^2 + (0.89)(0.36)s + (0.36)^2}, \quad \frac{v_c}{\delta_s} = 0;$$

$$\frac{q_c}{\delta_s} = \frac{35.12(s+0.5)}{s^2 + 2(0.89)(2/24)s + (2.24)^2}, \quad \frac{q_c}{v_{sel}} = 0 \quad (4.3)$$

v_c and q_c represents the desired velocity and pitch rate and these two trajectoried are shown in Fig.4.2, and δ_s is stick deflection. Physical constraints on the elevator are defined as $-26.5deg \leq \delta_{ea} \leq 6.75deg$.

2.2. Evolution of Off-Line Training

In the proposed indirect control strategy, fully tuned RBFN identifier using the Lyapunov-based stable tuning rule, as well as the MRAN's EKF tuning rule are trained off-line to learn the inverse dynamics of the aircraft. Then, in the control stage, both the neural networks (A and B) in Fig.4.1 are initialized with the off-line trained network parameters. In order to start with a comparatively compact network structure, a growing and pruning (GAP) strategy is adopted in implementing the RBFN structure.

Data pairs for the off-line training were generated from the open-loop aircraft dynamics, consisting of the actuator model and the aircraft model. In [89],

Indirect Adaptive Control Using Fully Tuned RBFN

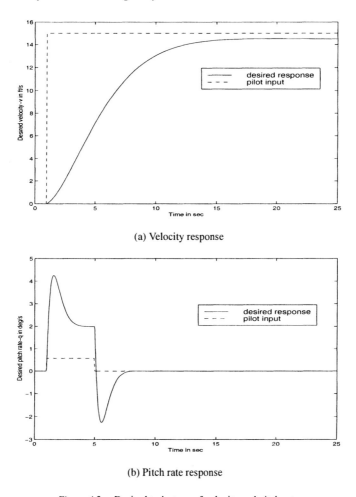

(a) Velocity response

(b) Pitch rate response

Figure 4.2. Desired trajectory of velocity and pitch rate

Nigel has studied this off-line training/on-line control strategy using MRAN algorithm. It was shown that with a GAP RBFN, the training pair used may affect the performance of the inverse neuro-controller when it is used in an on-line mode. To verify the results obtained in [89], the same scenarios as presented in [89] is utilized for off-line training in this section. However in this study we use the stable identification scheme as proposed in Chapter 3.

- **Case 1:** Training data pairs are arranged one after another in an appending manner.

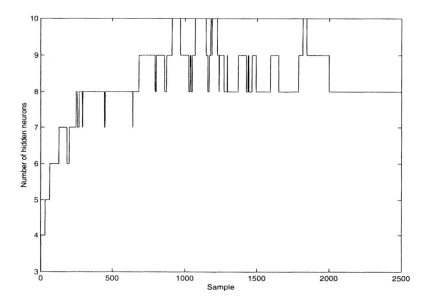

Figure 4.3. Neural history in off-line training

- **Case 2:** Training data pairs are arranged in an interleaven manner, whereby the sample of each data pair is interleavened to that of the other data pair.

- **Case 3:** Training data paris are arranged in a random manner, whereby the sample of each data pair is randomly arranged.

Prior to the formation of the training pair, the data pairs obtained from the open-loop dynamics are first normalized, which eliminates the possibility of any high magnitude input data (e.g. the velocity signal) overshadowing the smaller magnitude input data (e.g. the pitch rate signal) in contributing to the overall network output signal. Then the normalized data pairs are arranged in a selected manner (Case1, Case 2 or Case 3).

Although a stable identification scheme is utilized in the simulation studies, similar results are observed as shown in [89]: If the training data pairs are arranged as in Case 1 or Case 2, the off-line trained RBFN is unable to fully acquire the knowledge of the system dynamics. The reason is due to in the training procedure, the RBFN will grow neurons during the transient state of the training data pair, and prune neurons during the steady state of the training data pair. While compared to the first two cases, training the RBFN off-line using the randomly arranged training data pair provides the best command following

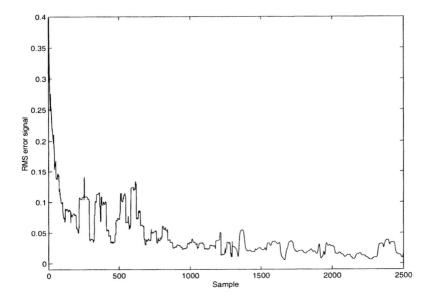

Figure 4.4. RMS error signal in off-line training

performance during the on-line simulation, since it can prevent unwanted loss of dynamics learned from the data pairs in the off-line training.

The training results using the stable identification scheme and the Lyapunov based tuning rule is presented in Fig.4.3 and Fig.4.4. Fig.4.3 shows the growing curve of the RBFN during the first sequence of the off-line training. It is seen that the number of neurons grows rapidly in the initial stage of the training, where the system dynamics of the aircraft model is acquired. Fig.4.4 presents the root-mean-square error signal during the off-line training and one can see it begin to converge after some number of samples. With this convergence, the growing number of neurons begin to settle into a steady state range of values whereby further training would not impose any significant change in the number of neurons.

Moreover, simulation studies of using the MRAN algorithm with EKF tuning rule is also carried out. It is observed that by using this tuning rule, although the approximation error is slightly small, it requires more calculation than using the Lyapunov-based tuning rule. Hence in real applications, a compromise has to be made between the accuracy and speed.

2.3. Off-Line Training/On-Line Control for the Linearized F8 Aircraft Model

After the off-line training has been completed, the resultant network's parameters are used to initialize the RBFN identifier (network B) and RBFN controller (network A) prior to on-line control. The main building blocks of this inverse control architecture comprise of the inverse model following control system and a neural network feedback component, the architecture together with all the signals are given in Fig.4.5. Network A takes the pilot input command signal $x_{des}(i + 1)$ as one of its input and the network B is fed with the actual output signal from the F8 aircraft model $x(i+1)$. Having different input signals fed to the two neural networks result in a difference between the outputs of the two networks ($\delta_c(i - 1)$ and $\delta'_c(i - 1)$). An error signal computed from this difference in the networks' output is used by the Lyapunov based tuning rule or MRAN algorithm in network B to adjust it's parameters. The newly calculated network B's parameters are then copied over into network A. The duplication takes place at the end of every sample for the whole flight simulation period.

In this study, during the on-line control stage, the performance of the three types of neural flight controller (NFC) using a fully tuned RBFN are investigated:

- **NFC-a:** RBFN with fixed number of hidden neurons.

- **NFC-b:** Growing RBFN (GRBFN) which can add the hidden neuron automatically.

- **NFC-c:** Growing and Pruning RBFN (GAP RBFN) which can add/prune the hidden neurons automatically.

Fig.4.6 presents the pilot command following results using the three methods respectively. It can be seen from the figure that using the NFC-a and NFC-b, a relatively good pilot's input command following performance is illustrated. However, NFC-c is indicating a divergence which is not acceptable.

The simulation studies indicate that RBFN controllers with fixed number or growing number of neurons are suitable for this off-line training/on-line control strategy. With reference to the control architecture, the output of network A represents the valid inverse aircraft model to the pilot's input command signals which results in the neuro-flight control system displaying remarkable tracking performance. As both the networks in the control architecture are a replication of one another, the true inverse aircraft model signal is then reproduced at the output of network B. The RBFN will learn on-line for any discrepancies in neuro-flight control system during the process of interpolation. However, for the NFC-c controller, with the GAP strategy, the learning algorithm has dynamically changed the network during its process of accomplishing its objective in reducing the error between the output signals of network A and B to minimal.

Indirect Adaptive Control Using Fully Tuned RBFN

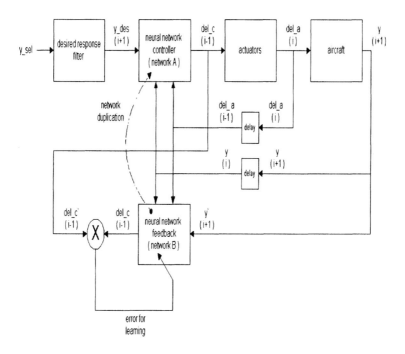

Figure 4.5. Exact inverse controller with full state feedback RBFN based flight control system

The results show that the neurons of the RBF network is pruned at the middle section of the on-line simulation. The resulting neuro-controller network therefore, is unable to perform the proper inverse approximation. It is therefore inadequate of using pruning strategy (NFC-c) for this off-line training/on-line control architecture.

3. On-Line Learning/On-Line Control for the Linearized Aircraft Model

The main disadvantage that lies in the off-line training/on-line control scheme is that an off-line training is a must before the control strategy is carried out [128]. To overcome this problem, another approach is that of the on-line learning/on-line control strategy. On-line learning/on-line control means that network B learns the inverse dynamics of the system in real-time, and network A is a duplication of network B, which is used to generate the desired control signal. However, in the literature, though there are a number of papers focusing

78   FULLY TUNED RBF NEURAL NETWORK FOR FLIGHT CONTROL

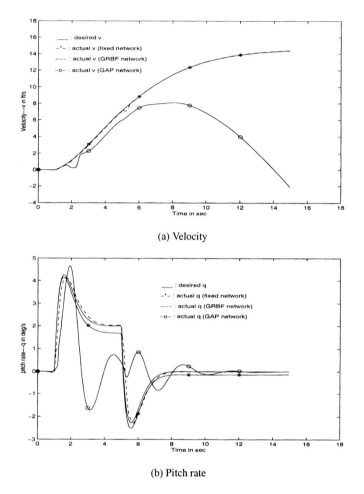

Figure 4.6. Performance evolution using off-line training/on-line control strategy

on the off-line training/on-line control strategies, for example, [85], [105], [7], etc., seldom any successful applications can be found using the indirect control strategy only in the on-line mode.

In this section, the performance of this strategy is explored based on the same F8 aircraft model. The performance of the three types of RBFN controller, NFC-a, NFC-b and NFC-c, are evaluated. Unfortunately, in this study, divergence is observed not only using the NFC-c controller, but also using the NFC-a and NFC-b strategy. The studies indicate that even using the RBFN with fixed number of hidden neurons, the performance for the on-line learning/on-line control scheme is poor without the prior knowledge of the system to be controlled.

In summary, in the indirect adaptive control case, the success of the control depends on the accuracy of the identification. When there is no off-line training, at the initial period, the network B can not exactly represent the inverse dynamics of the system, and hence network A outputs the wrong control signal. This wrong signal in turn results an incorrect output. In this way, the network B may learn the inverse dynamics of system in a region that outside its desired input-output profile. Therefore, the off-line training was essential in the indirect control architecture because the inverse neural controller was unable to adapt to flight condition which was not learned earlier, resulting in poor flight tracking performance, even divergence of the overall system.

It can be seen that the indirect adaptive control using RBFN suffers from several disadvantages:

- An off-line training is essential for acquiring the dynamics of the system inverse.
- Pruning strategy is not suitable for this control structure.
- A strict mathematical proof to guarantee the stability of the overall system and the convergence of the network parameters is still under exploration.

In this context, further studies on the neural controller design using fully tuned RBFN are needed.

4. Summary

This chapter emphasizes the evaluation of indirect adaptive control using fully tuned RBFN. The results obtained can be summarized as follows.

- Indirect adaptive control can be classified into off-line training/on-line control and on-line learning/on-line control strategies. Simulation studies indicate that using on-line learning/on-line control strategy, the performance is poor due to lack of prior knowledge.
- For off-line training/on-line control, it is shown that to implement a compact network structure, the adding/pruning strategy should be adopted in off-line training, and the randomly arranged training data pair provides the best command following performance during on-line simulation.
- In on-line control stage for off-line training/on-line control, the NFC-a controller (RBFN with fixed number of hidden neurons) and NFC-b controller (RBFN with growing strategy) exhibit good performance. However, divergence is observed in using the NFC-c controller (RBFN with growing/pruning strategy), which indicates a pruning strategy is not suitable for this control architecture.

In spite of the limitations, the off-line training/on-line control strategy does provide some valuable insights to practical application and research in the field of intelligent control design. However, to overcome the problems that exist in the indirect adaptive control scheme, the direct adaptive control scheme is studied in Part II.

II

DIRECT ADAPTIVE CONTROL STRATEGY AND FIGHTER AIRCRAFT APPLICATIONS

In order to overcome the drawbacks seen in the indirect adaptive control strategy, the RBFN based direct adaptive control strategy is explored in Part II, and forms a major contribution in this book.

Part II consists of three chapters. In Chapter 6, based on the feedback-error-learning control architecture, the parameter tuning rule for a fully tuned RBFN controller is derived based on the Lyapunov synthesis approach, guaranteeing the stability of the overall system. In this strategy, the RBFN controller, consisting of variable Gaussian functions, uses only on-line learning to represent the inverse dynamics of the aircraft and no off-line training is needed. This chapter also contributes to the robustness analysis of the proposed scheme in the presence of both the network approximation error and system model errors.

Chapter 7 demonstrates the theoretical results obtained in Chapter 6 through a series of simulation studies for the aircraft flight control applications. The performances of the proposed direct adaptive control scheme are evaluated for controlling the aircraft maneuver based on different aircraft models and control objectives. They vary from a linearized longitudinal F8 aircraft model for command following, designing controller to follow desired pitch rate signal based on a localized nonlinear F16 aircraft model, to maneuvering a full-fledged 6-DOF high performance aircraft to perform a high α stability-axis roll rate which is demonstrated to be highly nonlinear.

In Chapter 8, the performance of the newly proposed MRAN algorithm is explored for the first time in implementing a robust controller for the linearized longitudinal F8 aircraft model. Simulation results show that with MRAN's on-line learning ability, the proposed neuro-controller can tolerate aircraft model errors even up to 70%.

Chapter 5

DIRECT ADAPTIVE NEURO FLIGHT CONTROLLER USING FULLY TUNED RBFN

1. Overview

Control laws and design methods incorporating ANNs have been intensively studied in the area of aircraft flight control. In [17], Calise *et al.* have summarized some current research efforts of applying NN technology for flight control system design, with emphasis on nonlinear adaptive control. It has been shown that NN with on-line learning can adapt to aircraft dynamics which is poorly known or rapidly changing. However, in most of these applications, feedforward network with BP learning algorithm or its extensions has been the main paradigm, and there are only limited papers which explore the application of RBFN.

In [113], Singh *et al.* used a RBFN to suppress wing rock for a slender delta wing configuration. Calise presented the use of RBFN to capture variations in Mach number in [15], as these variations are difficult to be represented by polynomic functions in the transonic region. For on-line implementation, Sanner and Slotine [107] developed a direct adaptive tracking control architecture using Gaussian RBFN to adaptively compensate for plant nonlinearities in a robotic manipulator. Gomi and Kawato [27] also proposed a "feedback-error-learning" control strategy, where a Gaussian RBFN is used for on-line learning the inverse dynamics of a robotic system. In both schemes, the adaptive tuning rules are derived using Lyapunov synthesis approach, which guarantees closed-loop stability.

However, in all the above applications, the network structure is implemented using the classical approach, where only the weights of the networks are adjusted. As discussed in the previous chapters, a RBFN with all the parameters being updated can capture the system dynamics more quickly and accurately, and hence is more suitable for aircraft flight control. It is in this context that we attempt to explore the use of the fully tuned RBFN for aircraft control applications. An on-line control architecture originating from Kawato's feedback-error-learning scheme [27] is proposed in this chapter, which incorporates a fully tuned RBFN controller. Using the Lyapunov synthesis approach, a stable tuning rule for adjusting all the parameters of this RBFN is derived, guaranteeing the stability of the overall system. The robustness of the proposed tuning law in the presence of the neural network approximation error as well as the system model error is also analyzed. Although the derivation of this controller is similar to that of the stable nonlinear system identification in Chapter 3, they are definitely different in the actual mechanisms.

The rest of this chapter is organized as follows: Section 5.2 presents the problem formulation. Section 5.3 derives the stable adaptive tuning rule for a fully tuned RBFN. A robustness analysis of the proposed scheme to the approximation error and model error is presented in Section 5.4. In section 5.5, the derived tuning rule is implemented in a simple form under the discrete time framework.

2. Problem Formulation

The aircraft dynamics is represented by a continuous system,

$$\Sigma: \quad \dot{\mathbf{x}} = \mathbf{f}(\mathbf{x}, \mathbf{u}) \tag{5.1}$$

In Eq.(5.1), smoothness of $\mathbf{f}()$ is assumed. $\mathbf{x}$ is a bounded $n \times 1$ state vector, $\mathbf{u}$ is a bounded $p \times 1$ control vector. Without loss of generality, it is assumed that $\mathbf{f}(0,0) = 0$, so that the origin is an equilibrium state. In general, the number of control inputs is less than that of the states ($p < n$). It is known that under this condition only the p states can be tracked perfectly. Partitioning $\mathbf{x}$ into $\mathbf{x}_t$ and $\mathbf{x}_r$, the aircraft dynamics can be written as,

$$\Sigma: \quad \begin{pmatrix} \dot{\mathbf{x}}_t \\ \dot{\mathbf{x}}_r \end{pmatrix} = \begin{pmatrix} \mathbf{f}_t(\mathbf{x}, \mathbf{u}) \\ \mathbf{f}_r(\mathbf{x}, \mathbf{u}) \end{pmatrix} \tag{5.2}$$

The problem studied is to design a neuro-controller such that the given p states $\mathbf{x}_t$ of the aircraft can track the desired commands $\mathbf{x}_{dt}$ accurately, while at the same time the other states $\mathbf{x}_r$ asymptotically approach a equilibrium point $\mathbf{x}_{dr}$.

Determination of the conditions under which a solution $\mathbf{u}_d(t)$ exists forms part of the theoretical control problem. Based on the earlier studies in [84], to accomplish the above control objective, the system should satisfy Assumption 1.

Assumption 1: $\mathbf{f}_t(\mathbf{x}, \mathbf{u})$ has continuous bounded partial derivatives in a certain neighborhood of all the points along the desired trajectory $\mathbf{x}_d$ and the matrix $\frac{\partial \mathbf{f}_t(\mathbf{x},\mathbf{u})}{\partial \mathbf{u}^T}$ is nonsingular.

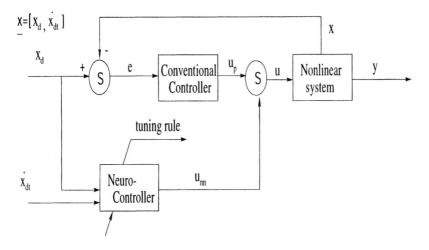

Figure 5.1. Control structure of neuro-flight-controller

It follows from the Implicit Function Theorem that the desired $\mathbf{u}_d(t)$ can be expressed as,

$$\mathbf{u}_d(t) = \bar{\mathbf{f}}_t(\mathbf{x}_d, \dot{\mathbf{x}}_{dt}) \qquad (5.3)$$

where $\bar{\mathbf{f}}_t$, a $p \times 1$ smooth function, is the inverse function of $\mathbf{f}_t$, and $\mathbf{x}_d = [\mathbf{x}_{dt}^T, \mathbf{x}_{dr}^T]^T$.

In the next section a RBFN controller will be utilized to approximate Eq.(5.3), and the tuning rule for updating all the parameters of the RBFN is derived according to a Lyapunov synthesis approach.

3. Stable Tuning Rule Using Fully Tuned RBFN

3.1. Control Strategy

The tuning law for adapting the weights of the RBFN has been derived in earlier papers [107][27], while in this chapter, the updating rule for a fully tuned RBFN controller is derived based on a feedback-error-learning strategy.

An on-line control strategy, which is similar to the well-known "feedback-error-learning" scheme, is proposed in Fig.5.1. This scheme is adopted because it has the advantage of generating the desired input signal without requiring that the network be trained initially off-line. In this case, the outputs of the neuro-controller $\mathbf{u}_{nn}$, which has fault tolerant ability through on-line learning, only depend on the desired input profile.

As mentioned above, $\mathbf{x}$ has been divided into the p states $\mathbf{x}_t$ that should track $\mathbf{x}_{dt}$ accurately, and the $n - p$ states $\mathbf{x}_r$ that should asymptotically approach a equilibrium point x_{dr} at the ending of the desired maneuver. Assuming all the states of the system are accessible, the tracking error e is defined as $e = x - x_d$, where $x_d = [x_{dt}{}^T, x_{dr}{}^T]^T$. This partition is used for analyzing the condition of the existence of the solution. In the following sections, they are combined together to derive the tuning rule.

The error dynamics for the overall system is,

$$\dot{e} = \dot{x} - \dot{x}_d = f(x, u) - f(x_d, u_d) \tag{5.4}$$

Using Taylor series expansion,

$$\begin{aligned}\dot{e} &= \frac{\partial f(x, u)}{\partial x^T}\bigg|_{x_d, u_d}(x - x_d) + o(x - x_d) \\ &+ \frac{\partial f(x, u)}{\partial u^T}\bigg|_{x_d, u_d}(u - u_d) + o(u - u_d)\end{aligned} \tag{5.5}$$

where $o()$ represents higher order terms. Substituting $\partial f(x, u)/\partial x^T|_{x_d, u_d}$ by $A(t)$, and $\partial f(x, u)/\partial u^T|_{x_d, u_d}$ by $B(t)$, neglecting all the higher order terms,

$$\dot{e} \approx A(t)e + B(t)(u - u_d) \tag{5.6}$$

With the widely used technology of the Control Configured Vehicle (CCV), the basic airframe may be designed to have low or even negative static stability in certain flight regimes, the augmented stability is then implemented by considering the controller's dynamics. In this study, a conventional controller is used in the strategy to improve the stability and response of the closed-loop system. This conventional controller can be realized by any approach like a pole-placement, a Linear Quadratic Regulator (LQR), or a H_∞ controller. To illustrate the concept, a proportional controller is designed which satisfies Assumption 2.

Assumption 2: The closed loop dynamic system, whose feedback controller is designed based on the nominal aircraft model, is stable along its desired flight trajectory.

With only the linear proportional controller $u = K_p(t)e$, the error dynamics is,

$$\dot{e} = (A(t) + B(t)K_p(t))e - B(t)u_d \tag{5.7}$$

According to Assumption 2, $K_p(t)$ is properly designed so that $J(t) = A(t) + B(t)K_p(t)$ is stable. Thus the total control signal to the aircraft is the sum of the proportional controller and the RBFN controller signals,

$$u = K_p(t)e + u_{nn} \tag{5.8}$$

3.2. RBFN Approximation and Error Dynamics

Setting the RBFN's inputs $\boldsymbol{\xi}$ to $\boldsymbol{\xi} = [\mathbf{x_d}^T, \dot{\mathbf{x}}_{\mathbf{dt}}^T]^T$, and using the well known universal approximation theory [92], $\mathbf{u_d}$ can be approximated by a RBFN controller through on-line learning,

$$\begin{aligned}
\mathbf{u_d} &= \mathbf{u}_{nn}^* + \epsilon_h \\
&= \sum_{k=1}^{h} \underline{\mathbf{w}}_k^{*T} \exp(-\frac{1}{\sigma_k^{*2}}\|\boldsymbol{\xi} - \boldsymbol{\mu}_k^*\|^2) + \epsilon_h \\
&= \mathbf{W}^{*T}\boldsymbol{\phi}(\boldsymbol{\mu}^*, \sigma^*, \boldsymbol{\xi}) + \epsilon_h
\end{aligned} \quad (5.9)$$

where $\mathbf{W}^*$ is $h \times p$ optimal weight matrix (h indicates the number of hidden neurons), $\underline{\mathbf{w}}_k^*$ is the kth row of $\mathbf{W}^*$, and $\boldsymbol{\phi}$ is $h \times 1$ Gaussian function, which is determined by the optimal centers $\boldsymbol{\mu}^*$ and widths σ^*. Using approximation theory, the inherent approximation error ϵ_h can be reduced arbitrarily by increasing the number of hidden neurons h [92]. Then it is reasonable to assume that ϵ_h is bounded by a constant ϵ_H, and

$$\epsilon_H = \sup_{\mathbf{x_d} \in \overline{\mathbf{X}}} \|\epsilon_h(\mathbf{x_d}, \dot{\mathbf{x}}_{\mathbf{dt}})\| \quad (5.10)$$

$\overline{\mathbf{X}}$ is the state space of the system. $\mathbf{x_d}$ and $\dot{\mathbf{x}}_{\mathbf{dt}}$ are assumed to be in the compact sets.

With the RBFN controller, from Eq.(5.8), the control input vector $\mathbf{u}$ is,

$$\mathbf{u} = \sum_{k=1}^{h} \hat{\underline{\mathbf{w}}}_k^T \exp(-\frac{1}{\hat{\sigma}_k^2}\|\boldsymbol{\xi} - \hat{\boldsymbol{\mu}}_k\|^2) + \mathbf{K_p}(t)\mathbf{e} = \hat{\mathbf{W}}^T\hat{\boldsymbol{\phi}} + \mathbf{K_p}(t)\mathbf{e} \quad (5.11)$$

where $\hat{\underline{\mathbf{w}}}_\mathbf{k}$ is the estimated weights, $\hat{\boldsymbol{\mu}}_k$ and $\hat{\sigma}_k$ are the estimated center and width for the kth hidden neuron. $\hat{\mathbf{W}}$ and $\hat{\boldsymbol{\phi}}$ are the corresponding matrix and vector expressions. Substituting Eq.(5.8), Eq.(5.11) into Eq.(5.6), the error dynamics becomes,

$$\dot{\mathbf{e}} = (\mathbf{A}(t) + \mathbf{B}(t)\mathbf{K_p}(t))\mathbf{e} + \mathbf{B}(t)(\hat{\mathbf{W}}^T\hat{\boldsymbol{\phi}} - \mathbf{W}^{*T}\boldsymbol{\phi}^* - \epsilon_h) \quad (5.12)$$

Using $\mathbf{J}(t) = \mathbf{A}(t) + \mathbf{B}(t)\mathbf{K_p}(t)$, defining $\tilde{\mathbf{W}}$ the difference between $\mathbf{W}^*$ and $\hat{\mathbf{W}}$, i.e. $\tilde{\mathbf{W}} = \mathbf{W}^* - \hat{\mathbf{W}}$, $\tilde{\boldsymbol{\phi}}$ the difference between $\boldsymbol{\phi}^*$ and $\hat{\boldsymbol{\phi}}$, i.e. $\tilde{\boldsymbol{\phi}} = \boldsymbol{\phi}^* - \hat{\boldsymbol{\phi}}$, the error dynamics may be written as,

$$\begin{aligned}
\dot{\mathbf{e}} &= \mathbf{J}(t)\mathbf{e} - \mathbf{B}(t)(\hat{\mathbf{W}}^T\tilde{\boldsymbol{\phi}} + \tilde{\mathbf{W}}^T\hat{\boldsymbol{\phi}} + \tilde{\mathbf{W}}^T\tilde{\boldsymbol{\phi}}) - \mathbf{B}(t)\epsilon_h \\
&\approx \mathbf{J}(t)\mathbf{e} - \mathbf{B}(t)(\hat{\mathbf{W}}^T\tilde{\boldsymbol{\phi}} + \tilde{\mathbf{W}}^T\hat{\boldsymbol{\phi}}) - \mathbf{B}(t)\epsilon_h
\end{aligned} \quad (5.13)$$

where $\mathbf{B}(t)(\hat{\mathbf{W}}^T\tilde{\boldsymbol{\phi}} + \tilde{\mathbf{W}}^T\hat{\boldsymbol{\phi}})$ represents the learning error E_l.

3.3. Stable Adaptive Tuning Rule for Fully Tuned RBFN

To derive the stable tuning law, choose the following Lyapunov candidate function,

$$V = \frac{1}{2}\mathbf{e}^T\mathbf{P}(t)\mathbf{e} + \frac{1}{2}tr(\tilde{\mathbf{W}}^T\Theta\tilde{\mathbf{W}}) + \frac{1}{2}\tilde{\phi}^T\Lambda\tilde{\phi} \quad (5.14)$$

where $\mathbf{P}(t)$ is an $n \times n$ time varying, symmetric, and positive definite matrix, and Θ, Λ are $h \times h$ non-negative definite matrices, the derivative of the Lyapunov function is given by,

$$\dot{V} = \frac{1}{2}[\dot{\mathbf{e}}^T\mathbf{P}(t)\mathbf{e} + \mathbf{e}^T\mathbf{P}(t)\dot{\mathbf{e}} + \mathbf{e}^T\dot{\mathbf{P}}(t)\mathbf{e}] + tr(\tilde{\mathbf{W}}^T\Theta\dot{\tilde{\mathbf{W}}}) + \tilde{\phi}^T\Lambda\dot{\tilde{\phi}} \quad (5.15)$$

Substituting Eq.(5.12) into Eq.(5.15),

$$\begin{aligned}\dot{V} =\ & -\mathbf{e}^T\mathbf{Q}(t)\mathbf{e} - \epsilon_h^T\mathbf{B}(t)^T\mathbf{P}(t)\mathbf{e} - \tilde{\phi}^T\hat{\mathbf{W}}\mathbf{B}(t)^T\mathbf{P}(t)\mathbf{e} \\ & - \hat{\phi}^T\tilde{\mathbf{W}}\mathbf{B}(t)^T\mathbf{P}(t)\mathbf{e} + tr(\tilde{\mathbf{W}}^T\Theta\dot{\tilde{\mathbf{W}}}) + \tilde{\phi}^T\Lambda\dot{\tilde{\phi}}\end{aligned} \quad (5.16)$$

where $\mathbf{Q}(t) = -\frac{1}{2}(\mathbf{J}(t)^T\mathbf{P}(t) + \mathbf{P}(t)\mathbf{J}(t) + \dot{\mathbf{P}}(t))$. Since $\mathbf{J}(t)$ is stable, the Lyapunov function can always be found and has a unique solution. Noting in Eq.(5.15),

$$tr(\tilde{\mathbf{W}}^T\Theta\dot{\tilde{\mathbf{W}}}) = \sum_{i=1}^{p}\tilde{\mathbf{w}}_i^T\Theta\dot{\tilde{\mathbf{w}}}_i \quad (5.17)$$

$$\hat{\phi}^T\tilde{\mathbf{W}}\mathbf{B}(t)^T\mathbf{P}(t)\mathbf{e} = \sum_{i=1}^{p}\hat{\phi}^T\tilde{\mathbf{w}}_i\mathbf{B}_i(t)^T\mathbf{P}(t)\mathbf{e} \quad (5.18)$$

Eq.(5.15) becomes,

$$\begin{aligned}\dot{V} =\ & -\mathbf{e}^T\mathbf{Q}(t)\mathbf{e} - \epsilon_h^T\mathbf{B}(t)^T\mathbf{P}(t)\mathbf{e} + \tilde{\phi}^T(-\hat{\mathbf{W}}\mathbf{B}(t)^T\mathbf{P}(t)\mathbf{e} + \Lambda\dot{\tilde{\phi}}) \\ & + \sum_{i=1}^{p}(-\tilde{\mathbf{w}}_i^T\hat{\phi}\mathbf{B}_i(t)^T\mathbf{P}(t)\mathbf{e} + \tilde{\mathbf{w}}_i^T\Theta\dot{\tilde{\mathbf{w}}}_i)\end{aligned} \quad (5.19)$$

where $\dot{\tilde{\mathbf{w}}}_i$ is the ith column of matrix $\dot{\tilde{\mathbf{W}}}$, $\mathbf{B}_i(t)$ is the ith column of matrix $\mathbf{B}(t)$.

If $\dot{\tilde{\mathbf{w}}}_i$ and $\dot{\tilde{\phi}}$ are selected as,

$$\dot{\tilde{\mathbf{w}}}_i = \Theta^{-1}\hat{\phi}\mathbf{B}(t)_i^T\mathbf{P}(t)\mathbf{e}, \quad i = 1,...,p \quad (5.20)$$

$$\dot{\tilde{\phi}} = \Lambda^{-1}\hat{\mathbf{W}}\mathbf{B}(t)^T\mathbf{P}(t)\mathbf{e} \quad (5.21)$$

Eq.(5.18) becomes,

$$\dot{V} = -\mathbf{e}^T\mathbf{Q}(t)\mathbf{e} - \epsilon_h^T\mathbf{B}(t)^T\mathbf{P}(t)\mathbf{e} \quad (5.22)$$

Direct Adaptive Neuro Flight Controller Using Fully Tuned RBFN

$\dot{V}$ can be demonstrated negative according to Eq.(5.10),

$$\dot{V} \leq -\|\mathbf{e}\|\lambda_{min}(\mathbf{Q})\|\mathbf{e}\| + \|\mathbf{e}\|\lambda_{max}(\mathbf{P})\,\|\mathbf{B}(t)\|\epsilon_H \quad (5.23)$$

Let $\|\mathbf{B}(t)\|\epsilon_H = \delta_{\epsilon_h}$, it can be derived directly that $\dot{V}$ is non-positive when

$$\|\mathbf{e}\| \geq \frac{\lambda_{max}(\mathbf{P})}{\lambda_{min}(\mathbf{Q})}\delta_{\epsilon_h} = E_a \quad (5.24)$$

Using the universal approximation proposition [92], by increasing the number h, ϵ_H can be reduced arbitrarily small, which means that E_a tends to zero when $h \to \infty$ and the negativeness of the Lyapunov function can be guaranteed, resulting in the overall system to be stable along the desired trajectory. However, it should be noted that in the real implementation, $h \to \infty$ is impossible, in this case the uniform ultimate boundedness (UUB) of the error signals is achieved.

Since $\dot{\tilde{\mathbf{w}}}_i = \dot{\mathbf{w}}_i^* - \dot{\hat{\mathbf{w}}}_i$, $\dot{\tilde{\phi}} = \dot{\phi}^* - \dot{\hat{\phi}}$ and $\dot{\mathbf{w}}_i^* = \mathbf{0}$, $\dot{\phi}^* = \mathbf{0}$, the tuning rules are,

$$\dot{\hat{\mathbf{w}}}_i = -\Theta^{-1}\hat{\phi}\mathbf{B}_i(t)^T\mathbf{P}(t)\mathbf{e}, \quad i = 1, ..., p \quad (5.25)$$

$$\dot{\hat{\phi}} = -\Lambda^{-1}\hat{\mathbf{W}}\mathbf{B}(t)^T\mathbf{P}(t)\mathbf{e} \quad (5.26)$$

4. Robustness Analysis

It can be seen that using the proposed method, provided Assumption 2 stands, the nonlinear function $\mathbf{f}()$ need not be known exactly. On the other hand, since the tuning rules derived involve the prior knowledge of matrix $\mathbf{B}(t)$, the robustness of the tuning laws is analyzed in case $\mathbf{B}(t)$ is partially known or unknown.

Case 1: Matrix $\mathbf{B}(t)$ is varying, but satisfies $\mathbf{B}(t) = k(t)\mathbf{B}_0$, where $\mathbf{B}_0$ is a known nominal value. $k(t)$ is a positive scalar varying with time.

Provided the weight tuning rule and the tuning rule of ϕ are given separately as,

$$\dot{\hat{\mathbf{w}}}_i = -k(t)\Theta^{-1}\hat{\phi}\mathbf{B}_{0i}^T\mathbf{P}(t)\mathbf{e}, \quad i = 1, ..., p \quad (5.27)$$

$$\dot{\hat{\phi}} = -k(t)\Lambda^{-1}\hat{\mathbf{W}}\mathbf{B}_0^T\mathbf{P}(t)\mathbf{e} \quad (5.28)$$

$\dot{V}$ can be demonstrated to be negative using the same condition as in Eq.(5.24). In implementation, $k(t)$ can be replaced by a positive scalar η, because η and $k(t)$ have the same sign, the system's convergence characteristic will not change.

Case 2: $\mathbf{B}(t)$ is unknown and it is represented by a nominal value plus its variation, $\mathbf{B}(t) = \mathbf{B}_0(\mathbf{I} + \Delta\mathbf{B}(t))$.

Assume $\|\Delta\mathbf{B}(t)\| \leq M$, M is the upper bound for the uncertainty. The derivative of Lyapunov function of Eq.(5.15) is re-analyzed and is given by,

$$\dot{V} = -\mathbf{e}^T\mathbf{Q}(t)\mathbf{e} - \mathbf{e}^T\mathbf{P}(t)(\mathbf{B}_0 + \mathbf{B}_0\Delta\mathbf{B}(t))\boldsymbol{\epsilon} - (\tilde{\phi}^T\hat{\mathbf{W}}\mathbf{B}_0\Delta\mathbf{B}(t)\mathbf{P}(t)\mathbf{e}$$

$$+\hat{\phi}^T\tilde{\mathbf{W}}\mathbf{B}_0\Delta\mathbf{B}(t)\mathbf{P}(t)\mathbf{e}) + \tilde{\phi}^T(-\hat{\mathbf{W}}(\mathbf{B}_0 + \mathbf{B}_0\Delta\mathbf{B}(t))^T\mathbf{P}(t)\mathbf{e} + \Lambda\dot{\tilde{\phi}})$$
$$+ \sum_{i=1}^{p}(-\underline{\tilde{\mathbf{w}}}_i^T\hat{\phi}(\mathbf{B}(t)_{0i} + \mathbf{B}(t)_{0i}\Delta\mathbf{B}(t)_i)^T\mathbf{P}(t)\mathbf{e} + \underline{\tilde{\mathbf{w}}}_i^T\Theta\dot{\tilde{\mathbf{w}}}_i) \qquad (5.29)$$

Re-writing the weight tuning rule and the tuning rule of ϕ using $\mathbf{B}_0$,

$$\dot{\hat{\mathbf{w}}}_i = -\Theta^{-1}\hat{\phi}\mathbf{B}_{0i}^T\mathbf{P}(t)\mathbf{e}, \quad i = 1,...,p \qquad (5.30)$$
$$\dot{\hat{\phi}} = -\Lambda^{-1}\hat{\mathbf{W}}\mathbf{B}_0^T\mathbf{P}(t)\mathbf{e} \qquad (5.31)$$

then,

$$\dot{V} = -\mathbf{e}^T\mathbf{Q}\mathbf{e} - \mathbf{e}^T\mathbf{P}(\mathbf{B}_0 + \mathbf{B}_0\Delta\mathbf{B}(t))\epsilon_h$$
$$-(\tilde{\phi}^T\hat{\mathbf{W}} + \hat{\phi}^T\tilde{\mathbf{W}})\mathbf{B}_0\Delta\mathbf{B}(t)\mathbf{P}\mathbf{e} \qquad (5.32)$$

Re-analyzing $\dot{V}$, we have,

$$\dot{V} \leq -\|\mathbf{e}\|\lambda_{min}(\mathbf{Q})\|\mathbf{e}\| + \|\mathbf{e}\|\lambda_{max}(\mathbf{P})\|\mathbf{B}_0\|\|\mathbf{I} + \Delta\mathbf{B}(t)\|\epsilon_H + \|\tilde{\phi}^T\hat{\mathbf{W}}$$
$$+\hat{\phi}^T\tilde{\mathbf{W}}\|\|\mathbf{B}_0\|\|\Delta\mathbf{B}(t)\|\lambda_{max}(\mathbf{P})\|\mathbf{e}\| \qquad (5.33)$$

Since $\|\mathbf{I} + \Delta\mathbf{B}(t)\|\epsilon_H \leq \delta'_{\epsilon_h}$ and defining $\|(\tilde{\phi}^T\hat{\mathbf{W}} + \hat{\phi}^T\tilde{\mathbf{W}})\|\lambda_{max}(\mathbf{P}) = \xi(\tilde{\phi}, \tilde{\mathbf{W}})$ for brief, it can be shown directly that $\dot{V}$ is nonpositive when

$$\|\mathbf{e}\| \geq \frac{\lambda_{max}(\mathbf{P})\|\mathbf{B}_0\|}{\lambda_{min}(\mathbf{Q})}\delta'_{\epsilon_h} + \frac{\lambda_{max}(\mathbf{P})\|\mathbf{B}_0\|}{\lambda_{min}(\mathbf{Q})}\xi(\tilde{\phi}, \tilde{\mathbf{W}})\|\Delta\mathbf{B}(t)\|$$
$$= E_a + E_{lm} \qquad (5.34)$$

In the above equation, E_a is caused by the approximation inaccuracy ϵ_h, E_{lm} is caused by the product of the learning error $E_l(\|(\tilde{\phi}^T\hat{\mathbf{W}} + \hat{\phi}^T\tilde{\mathbf{W}})\|)$ and modeling error $E_m(\|\Delta\mathbf{B}(t)\|)$.

Remarks:

1 If there is neither an approximation error (*i.e.* $\epsilon_h = 0$) nor any model error (*i.e.* $\Delta\mathbf{B}(t) = 0$), from Eq.(5.32), $\dot{V}$ is negative semidefinite and hence the stability of the overall scheme is guaranteed.

2 The approximation error E_a is related to the number of hidden neurons used in the RBFN. Given enough hidden units, it will converge to a very small number.

3 The second error item E_{lm} is a combination of the learning error and modeling error. The product of ξ and $\Delta\mathbf{B}(t)$ indicates that even under large model error $\Delta\mathbf{B}(t)$, if the RBFN learns the desired value $\mathbf{W}^*$ and ϕ^* correctly, E_{lm} is still very small.

5. Implementation of the Tuning Rule

In Eq.(5.26), the Gaussian function $\hat{\phi}$ is embeded with the centers' locations $\hat{\mu}$ and widths $\hat{\sigma}$, i.e. $\hat{\phi} = \phi(\hat{\mu}, \hat{\sigma}, \xi)$. Combining all the adaptable parameters into a big vector, $\chi = [\underline{\hat{w}_1}, \hat{\mu}_1^T, \hat{\sigma}_1, \cdots, \underline{\hat{w}_h}, \hat{\mu}_h^T, \hat{\sigma}_h]^T$, a simple updating rule for χ can be derived. Because the real implementation is carried out in a discrete-time framework, the updating laws is also derived under a discrete form.

First, Eq.(5.25) can be converted into,

$$\dot{\hat{w}}_i = -\Theta^{-1}\hat{\phi}B_i(t)^T P(t)e, \quad i = 1,...,p$$
$$\Rightarrow \dot{\hat{W}}^T = -B(t)^T P(t)e\hat{\phi}^T \Rightarrow \dot{\underline{\hat{w}}}_k^T = -B(t)^T P(t)e\hat{\phi}_k,$$
$$k = 1,...,h \quad \Theta = I \quad (5.35)$$

where $\underline{\dot{\hat{w}}}_k$ is the kth row of matrix $\dot{\hat{W}}$. Define $\hat{g} = \hat{W}^T \hat{\phi}$ the output of the RBFN, $\hat{\phi}_k$ is the derivative of $\hat{g}()$ to the weights $\hat{w}_k$, this equation can be converted into a discrete form,

$$\underline{\hat{w}}_k^T(n+1) = \underline{\hat{w}}_k^T(n) - \tau \frac{\partial \hat{g}}{\partial \underline{\hat{w}}_k^T} B(n)^T P(n)e(n) \quad k = 1...h \quad (5.36)$$

where τ is the sampling time.

To derive the discrete tuning rules for the center and widths from Eq.5.26, we are using the same approach as introduced in section 3.2.3. Therefore, the updating laws for centers and widths of the basis functions are,

$$\hat{\mu}_k(n+1) = \hat{\mu}_k(n) - \tau \eta_1 \frac{\partial \hat{g}}{\partial \hat{\mu}_k^T} B(n)^T P(n)e(n)$$
$$k = 1,...,h \quad (5.37)$$

$$\hat{\sigma}_k(n+1) = \hat{\sigma}_k(n) - \tau \eta_2 \frac{\partial \hat{g}}{\partial \hat{\sigma}_k^T} B(n)^T P(n)e(n)$$
$$k = 1,...,h \quad (5.38)$$

In the above equations, η_1 and η_2 are positive scalars to be selected properly. Integrating Eq.(5.36), Eq.(5.36) and Eq.(5.38) by using the vector χ, the updating rule is given by,

$$\chi(n+1) = \chi(n) - \eta \Pi(n) B(n)^T P(n)e(n) \quad (5.39)$$

where η is learning rate and $\eta < min(\tau, \eta_1, \eta_2)$, $\Pi(n) = \nabla_\chi \hat{g}(\xi_n)$ is the gradient of the function $\hat{g}()$ with respect to the parameter vector χ evaluated at

$\chi(n)$, and

$$\Pi(n) = [\hat{\phi}_1 I_{p \times p}; \hat{\phi}_1 \frac{2(\boldsymbol{\xi}_n - \hat{\boldsymbol{\mu}}_1)\hat{\mathbf{w}}_1}{\hat{\sigma}_1^2}; \hat{\phi}_1 \frac{2\hat{\mathbf{w}}_1}{\hat{\sigma}_1^3}\|\boldsymbol{\xi}_n - \hat{\boldsymbol{\mu}}_1\|^2; \cdots;$$
$$\hat{\phi}_h I_{p \times p}; \hat{\phi}_h \frac{2(\boldsymbol{\xi}_n - \hat{\boldsymbol{\mu}}_h)\hat{\mathbf{w}}_h}{\hat{\sigma}_h^2}; \hat{\phi}_h \frac{2\hat{\mathbf{w}}_h}{\hat{\sigma}_h^3}\|\boldsymbol{\xi}_n - \hat{\boldsymbol{\mu}}_h\|^2] \quad (5.40)$$

Since the number of hidden neurons of the RBFN is fixed in this scheme, the approximation error E_a is quite large at the initial learning period. If $\|e\| < E_a$, then it is possible that $\dot{V} > 0$, which implies that the parameters of the network may drift to infinity. A common way to avoid this problem and ensure the convergence of approximation error is to incorporate a dead zone in the tuning rules [27]. The tuning rule is given now as,

$$\chi(n+1) = \begin{cases} \chi(n) - \eta \Pi(n) \mathbf{B}^T \mathbf{P}(n)\mathbf{e}(n) & \text{if } \|e\| \geq e_0 \\ & \text{and} \|\mathbf{W}\| \leq h^{\frac{1}{2}} Z \\ \chi(n) & \text{otherwise} \end{cases} \quad (5.41)$$

In Eq.(5.41), Z is a positive scalar and $h^{\frac{1}{2}} Z$ is an upper bound on $\|\mathbf{W}\|$ (Euclidean norm of weight matrix), e_0 is selected as the required accuracy on the state error $\mathbf{e}$. According to Eq.(5.24), the size of the dead zone should be set to E_a, so that if the error $\|e\| < E_a$, the tuning is stopped and the parameters will not drift away. Unfortunately, since ϵ_H is time-varying and unknown, the exact value for E_a can not be estimated, therefore e_0 is used in the place of $E_a(t)$. If $e_0 > E_a$, then $\dot{V}$ is always non-positive. If $e_0 < E_a$, when the error $\|e\|$ converges to $e_0 < \|e\| < E_a$, $\dot{V}$ may be positive and in this case the upper bound $h^{\frac{1}{2}} Z$ is used to prevent the parameter from drifting away.

6. Summary

In this chapter, by using a Lyapunov synthesis approach, a parameter tuning law is derived for a fully tuned RBFN controller, which can guarantee the stability of the overall system. A robustness analysis indicates that provided minor assumptions are satisfied and the inequality Eq.(5.33) stands, the performance of the proposed scheme is guaranteed even when the system dynamics is uncertain. In the next chapter, simulation studies based on different aircraft models are carried out to illustrate the derived parameter tuning rule and the performance of the proposed control architecture.

Chapter 6

AIRCRAFT FLIGHT CONTROL APPLICATIONS USING DIRECT ADAPTIVE NFC

This chapter explores the application of the direct adaptive neural control scheme incorporating the fully tuned RBFN controller proposed in Chapter 6 for aircraft control applications. To evaluate the performance of this neuro-flight-controller (NFC), the following scenarios are considered in this chapter.

- Tracking the desired velocity and pitch rate command based on a linearized longitudinal F8 aircraft model (Linearized model).

- Tracking the desired pitch rate command based on a localized longitudinal aircraft model (Localized nonlinearity).

- Tracking the desired trajectory in a high α stability-axis roll maneuver based on a full-fledged nonlinear 6 degree-of-freedom (DOF) high performance fighter aircraft model (full nonlinearity).

For comparison purposes, simulations using the fully tuned RBFN and the RBFN with only weights tuning are carried out.

1. Robust NFC for a Linearized F8 Aircraft Model

To start from the most simple case and also for comparison purpose, in this section, the proposed fully tuned RBFN controller is used for command following of a linearized longitudinal F8 aircraft model.

The linearized longitudinal F8 aircraft model has been presented in Section 5.2.1, where the model was utilized to test the performance of RBFN controller in the indirect adaptive control schemes. The same model is utilized here for demonstrating the performance of the proposed direct adaptive control. For clarity, the model and the control objective are recounted briefly.

In [105], Sadhukan and Feteih presented an exact inverse neuro-controller with full state feedback for a F8 aircraft. The control system is designed to provide a decoupled response to pilot velocity and pitch rate commands of the F8 aircraft model. The pilot inputs and the desired response as per the level 1 handling qualities requirements are shown in (Fig.6.1). The state vector of the model comprises of the velocity (v in ft/s), angle of attack (α in rad), pitch rate (q in rad/s) and pitch angle (θ in rad). The control inputs u include the elevator's output δ_{ea} in rad and the throttle's output δ_{ta} in percent. Physical constraints on the elevator are defined as $-26.5deg \leq \delta_{ea} \leq 6.75deg$. In this simulation, the test pilot inputs used for all the cases are a velocity (v) step command of magnitude $15ft/s$, and a pitch rate step command represented by a 0.5 in. of δ_s (stick deflection).

In [105], studies were carried out not only to evaluate the performance of the neuro-controller as that of Chapter 5, but also to evaluate the response in the presence of model errors. Under 40% model error, where all the items in both of the matrix $\mathbf{F}_1$ and $\mathbf{G}_1$ are varied by 40%, the velocity oscillates in the first few seconds though eventually settling, it does demonstrate the on-line learning ability of the neural network when the errors are small. However, with the traditional feedback controller and the RBF controller's on-line learning ability, it is shown that the proposed scheme can tolerate more serious fault conditions. To assess the performance of this neuro-flight-controller, the model errors are simulated by changing all the elements of the system matrix $\mathbf{F}_1$ and $\mathbf{G}_1$ by 70% for $t \geq 0s$.

For the open loop F8 linearized aircraft model, the roots of the characteristic equation are $-0.6656 \pm j2.81$ and $-0.0069 \pm j0.0765$. Though stable, the phugoid mode needs stability augmentation to achieve desired response characteristics like rise time, overshoot, etc. Using conventional controller design methods, a proportional controller is designed first as a baseline controller to provide stability augmentation. Actually, this controller can be realized by any other approach like pole-placement or Linear Quadratic Regulator (LQR) method. The proportional controller gain was selected as $\mathbf{K}_p = \begin{bmatrix} -0.01 & 0.002 & 2.92 & 0.02 \\ -2.02 & 0 & -0.12 & -0.30 \end{bmatrix}$, and good performance is observed in nominal condition from Fig.6.2. However, it can be seen that by using the traditional controller alone, in case of the 70% model error, the tracking results deteriorate greatly. Hence the RBFN controller is added in cascade with the aircraft, to improve the tracking accuracy by learning the inverse dynamics of the system.

At the start of the on-line learning, according to the *a priori* knowledge, two cases are studied where the number of hidden neurons were frozen at 6 and 15 separately, the centers of the hidden units are fixed with a grid method described in [110] and their widths are set to 0.5. Fig.6.3 plots the tracking performance under normal case, while Fig.6.4 shows the same under 70% model error. Fig.6.3 indicates that under the nominal condition, because of the exis-

Aircraft Flight Control Applications Using Direct Adaptive NFC

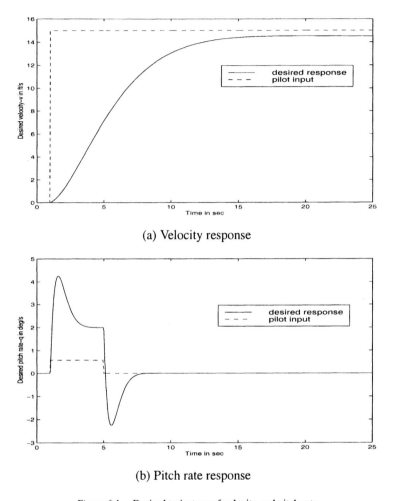

(a) Velocity response

(b) Pitch rate response

Figure 6.1. Desired trajectory of velocity and pitch rate

tence of the pre-designed feedback controller, all the strategies can achieve good performance. However, when there is 70% model error, better tracking result is observed using a fully tuned RBFN. From Fig.6.4 it can be easily seen that using fully tuned RBFN, the tracking performance is quite good even with only 6 hidden units, while if only the weights of the RBFN are updated, it is obvious that the controller can not track the desired commands correctly. Although increasing the number of hidden units to 15 improves the performance, it is still not as good as the RBFN with tuning all the parameters. Fig.6.5 depicts the

98 FULLY TUNED RBF NEURAL NETWORK FOR FLIGHT CONTROL

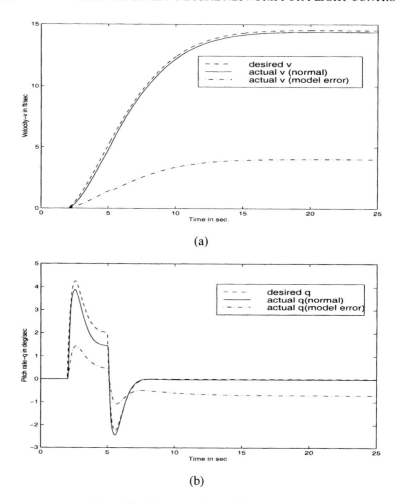

Figure 6.2. Response of the traditional controller

evolution of tracking errors of both velocity and pitch rate. A detailed quantitative comparison in terms of the average $E(ave)$ and peak tracking errors $E(max)$ for different sizes of the RBFNs are given in Table 6.1 (All the results are compared under 70% model error). This phenomena, in fact indicates that in using a traditional RBFN, *a priori* knowledge should be embedded into the centers and widths of the hidden units accurately. On the other hand, with a fully tuned RBFN, since its centers, widths can be modified during the on-line learning, this requirement is not so stringent.

Aircraft Flight Control Applications Using Direct Adaptive NFC

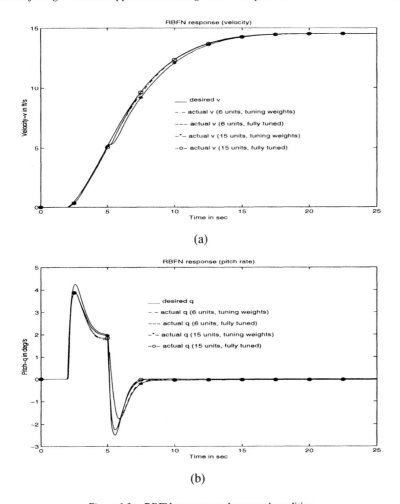

Figure 6.3. RBFN response under normal condition

Fig.6.6 shows how the inverse dynamics is learnt on-line by the RBFN controller. The continuous line in the figure represents the outputs of the RBFN controller, while the dotted line represents the conventional controller's outputs. Using a RBFN controller with only tuning the weights, the outputs of the two controllers under nominal condition and model error are given in the left column (Fig.6.6(a)) and middle column (Fig.6.6(b)) respectively. It can be seen from the figures that in the initial period, the conventional controller contributes a lot to maintain the performance, while the RBFN controller's outputs

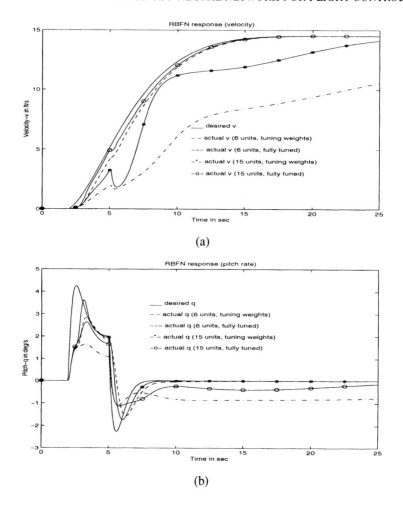

Figure 6.4. RBFN response under 70% model error

are varying, which indicates the procedure of learning. In nominal condition, after 10 seconds, the steady outputs are all due to the RBFN controller, and the outputs of the feedback controller shrinks to zero, which means that the RBFN controller has learnt the system inverse. However, under model error, the traditional RBFN fails to catch up with the inverse dynamics of the system. This is actually caused by the incorrect centers and widths used. Compared to Fig.6.6(b), it is shown in the right column (Fig.6.6(c)) that with a fully tuned RBFN, the changes in the model dynamics can be grasped quickly. As a result,

Aircraft Flight Control Applications Using Direct Adaptive NFC

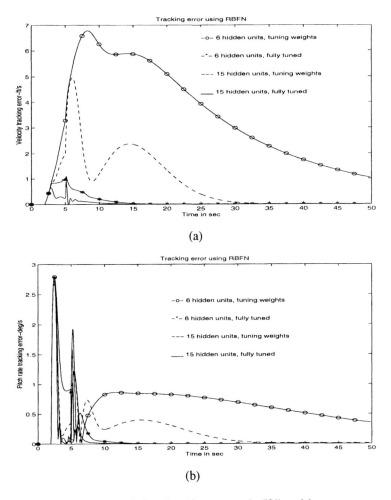

Figure 6.5. Evolution of tracking errors under 70% model error

the control inputs generated are mainly formed by the RBFN controller, and note that they are well within the limits without saturation.

Fig.6.7 plots the trajectories of the other two variables, namely, angle of attack and pitch angle and they are brought back to the trim points, which is guaranteed by the tuning laws derived using the Lyapunov stability theory. However, for the case of only tuning the weights, they can not be brought back to the trimmed points, as indicated by the dotted line.

Table 6.1. Tracking results comparison using RBFN (70% model error)

Network Structure	Adapted Parameters	Hidden Units	Tracking Error			
			$E_v m$	$E_v a$	$E_q m$	$E_q a$
RBFN	$\mathbf{W}$	6	6.78	3.37	2.79	0.68
RBFN	$\mathbf{W}, \mu, \theta$	6	1.02	0.11	2.78	0.09
RBFN	$\mathbf{W}$	15	4.96	0.83	2.67	0.22
RBFN	$\mathbf{W}, \mu, \theta$	15	0.92	0.042	2.63	0.062

Note: E_v in ft/s, E_q in deg/s.
$E_v m \equiv E_v(max), E_v a \equiv E_v(ave), E_q m \equiv E_q(max), E_q a \equiv E_q(ave)$

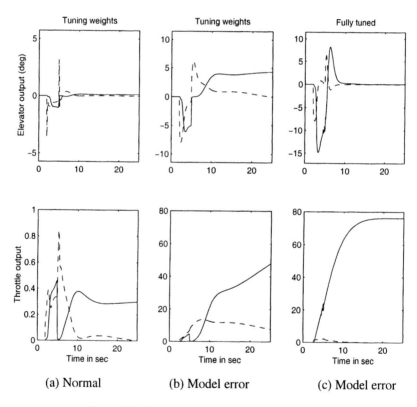

Figure 6.6. Control inputs (RBFN with 6 hidden units)

To implement a fully tuned RBFN, the number of the hidden neurons should be selected *a priori* by *ad hoc* or through a trial-and-error procedure. However, in case of approximating the inverse dynamics of a very complicated system,

Aircraft Flight Control Applications Using Direct Adaptive NFC 103

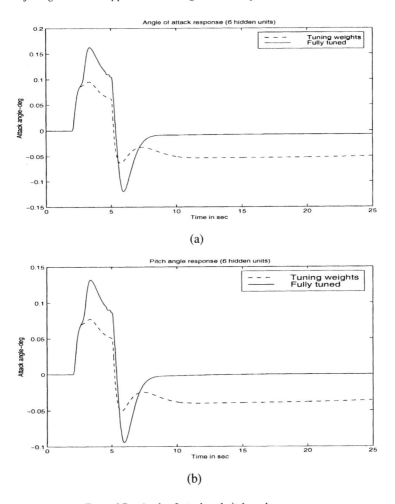

Figure 6.7. Angle of attack and pitch angle response

this procedure will waste a lot of time, and superfluous hidden neurons will be added to the network. To avoid this problem, the performance of dynamically structured RBFNs are investigated. The RBFN with a growing strategy is called a "GRBFN", and a RBFN with both growing and pruning strategy is known as a "GAP RBFN". The same growing and pruning strategy as described in [61][94] is utilized to implement the GRBFN/GAP RBFN controller. The simulations are intended to compare the tracking accuracy as well as the resulting network complexity using RBFN, GRBFN and GAP RBFN.

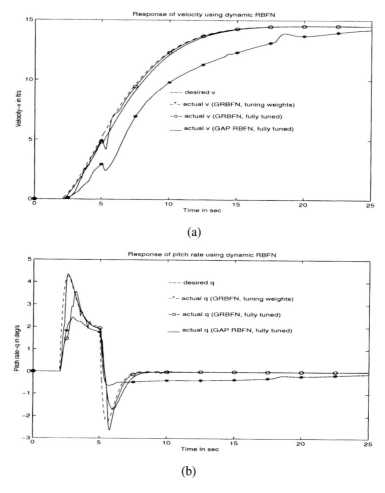

Figure 6.8. Dynamic structured RBFN response under model error

Fig.6.8 shows the tracking result under model error using a dynamic RBFN. Fig.6.9 represents the tracking error and Fig.6.10 describes the history for the hidden neurons. From the figures it can be seen the network gradually learns the system inverse by recruiting new hidden units according to the novelty of the input data. The results also illustrate that using the fully tuned strategy, the output error is much smaller than the RBFN with only tuning the weights (refer to Fig.6.9), and the network is more compact (less hidden neurons as indicated in Fig.6.10).

Aircraft Flight Control Applications Using Direct Adaptive NFC

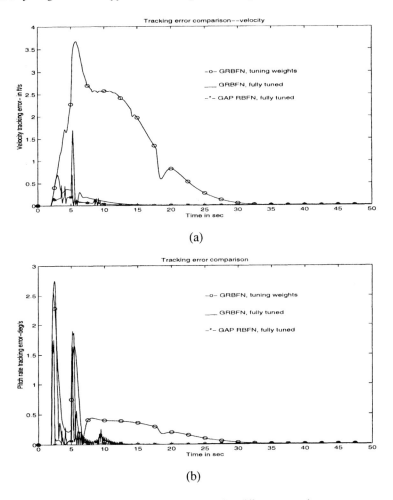

Figure 6.9. Evolution of tracking errors using different network structures

Quantitative comparisons are given in Table 6.2. It is shown in the table that using a GRBFN, with only tuning the weights, 24 hidden neurons are used, and the average tracking error for velocity and pitch rate are 0.79, 0.21 respectively, while with fully tuned GRBFN, these numbers decrease to 0.06 and 0.074, and the number of hidden units used is only 10. Compared with a fully tuned GRBFN, the GAP RBFN with a pruning strategy can even achieve better performance, that is, less hidden neurons (only 4 units are needed) and small tracking error ($E_v(ave) = 0.026, E_q(ave) = 0.037$).

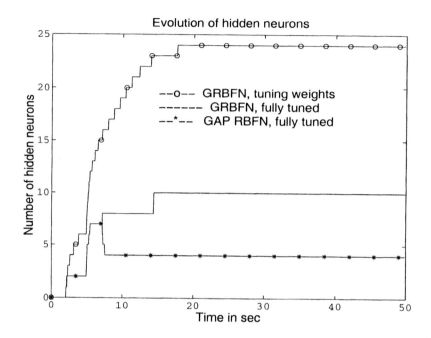

Figure 6.10. Evolution of hidden neurons using different network structures

Table 6.2. Tracking results comparison using different RBFN structures

Network Structure	Initialization	Adapted Parameters	Hidden Units	Tracking Error			
				$E_v m$	$E_v a$	$E_q m$	$E_q a$
RBFN	Grid	$\mathbf{W}$	15	4.96	0.83	2.67	0.22
RBFN	Grid	$\mathbf{W}, \mu, \theta$	15	0.92	0.042	2.63	0.062
GRBFN	no need	$\mathbf{W}$	24	3.67	0.79	2.75	0.21
GRBFN	no need	$\mathbf{W}, \mu, \theta$	10	1.69	0.06	2.55	0.074
GAP RBFN	no need	$\mathbf{W}, \mu, \theta$	4	0.69	0.026	1.75	0.037

Note: E_v in ft/s, E_q in deg/s.
$E_v m \equiv E_v(max), E_v a \equiv E_v(ave), E_q m \equiv E_q(max), E_q a \equiv E_q(ave)$

It should be pointed out that in the indirect adaptive control strategy, the divergence in the performance is observed if a pruning strategy is incorporated in realizing the network structure. However, it is shown in this simulation study that using this direct adaptive control strategy for controlling the linearized aircraft model, the network does not prune the hidden neuron in the steady state (Fig.6.10), and the performance is much better. Further studies on the evaluation of the pruning strategies in the direct adaptive control case is carried out in the next sections.

In summary, simulation studies based on the linearized longitudinal F8 aircraft model demonstrate that by using the proposed fully tuned NFC (RBFN controller), the performance is much better. Even under 70% model error, it can learn this uncertainty on-line and achieve a good command response. Finally, it worth noting that the study in this section is carried out for comparison purpose, the F8 model used is linearized under a given flight condition as that in [105]. However, under more realistic situations, the aircraft dynamics is actually a complicated nonlinear system, varying with the aerodynamic parameters, inertial properties and different flight conditions, which will be covered in the following two sections.

2. NFC for Localized Nonlinear Aircraft Model

The neuro-flight-controller (NFC) has been evaluated based on a linearized aircraft model in Section 6.1. In this section, further simulations are carried out based on a localized nonlinear longitudinal aircraft model as described in [121]. The objective is to control the aircraft to follow a given pitch rate command. The performance of the conventional controller as well as the NFC control strategy are analyzed under model error, which is simulated by increasing the nonlinear term of the aircraft model to a certain degree.

2.1. Localized Nonlinear Fighter Aircraft Model

The aircraft model is linearized at a medium velocity trim condition: $V_{trim} = 152m/s, \alpha_{trim} = 2.19°, \theta_{trim} = 0.03822rad, \delta_{htrim} = -0.5979°$, after linearization, the localized nonlinear aircraft dynamics at the above trim condition can be expressed as,

$$\dot{\mathbf{x}} = \mathbf{A}\mathbf{x} + \phi(\mathbf{x}) + \mathbf{B}\mathbf{u} \tag{6.1}$$

where $\mathbf{x} = [\triangle V, \triangle \alpha, \triangle \theta, \triangle q]^T$, $\mathbf{u}$ refers to the δ_{ea}, denoting the elevator's output in rad. The plant matrices $\mathbf{A}$, $\mathbf{B}$ and $\phi(\mathbf{x})$ in the model are given by,

$$\mathbf{A} = \begin{bmatrix} -0.01940 & 1.7666 & -9.80 & -0.1917 \\ -0.000845 & -1.002 & 0 & 0.9046 \\ 0 & 0 & 0 & 1 \\ 0.000426 & 0.7589 & 0 & -1.0765 \end{bmatrix} ; \quad \mathbf{B} = \begin{bmatrix} 0.03797 \\ -0.0024 \\ 0 \\ -0.1850 \end{bmatrix}$$

$$\phi(\mathbf{x}) = \begin{bmatrix} 0.0183x_1x_2 + 37.12x_2^2 - 0.0364x_1x_4 \\ -147.6x_2x_4 + 0.00257x_1x_2^2 - 0.01489x_2^3 \\ -0.2548x_2^2 - 0.0614x_2x_4 \\ +0.000466x_1x_2^2 - 0.00270x_2^2 \\ 0 \\ -0.001235x_1x_2 - 2.504x_2^2 - 0.0965x_2x_4 \\ +0.000506x_1x_2^2 - 0.00293x_2^3 \end{bmatrix}$$

2.2. Performance Evaluation of the NFC

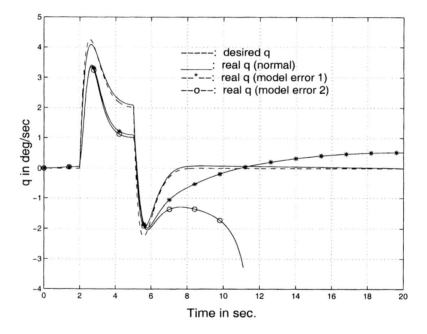

Figure 6.11. Simulation results using the conventional controller

At the beginning, a conventional controller is designed under the nominal condition. Fig.6.11 shows the results of using only the conventional controller (proportional controller). It can be seen from the figure that although the performance is acceptable when the aircraft model's nonlinear part is increased to 40 times of its original value (model error 1), a divergence is observed in case it is expanded to 60 times (model error 2). It is obvious that the stability of the overall system is related with the coefficient ζ which represents the local

Aircraft Flight Control Applications Using Direct Adaptive NFC

nonlinearity. For simplicity this phenomenon can be briefly explained based on a regulator problem.

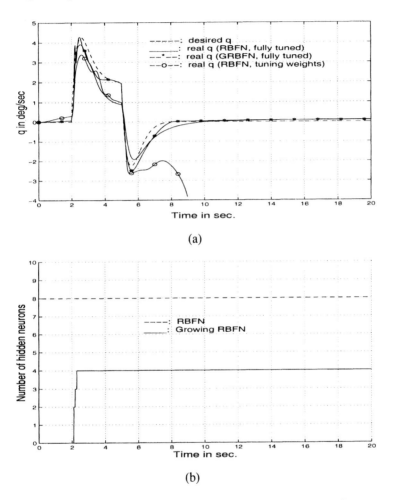

Figure 6.12. Simulation results using the NFC strategy (model error 2)

Consider the system: $\dot{\mathbf{x}} = \mathbf{f}(\mathbf{x}, \mathbf{u}) = \mathbf{A}\mathbf{x} + \zeta\mathbf{g}(\mathbf{x}) + \mathbf{B}\mathbf{u}$, where $\zeta\mathbf{g}(\mathbf{x})$ represents the local nonlinearity and $\mathbf{g}(\mathbf{x})$ is the higher-order term of $\mathbf{x}$. Let $V = \frac{1}{2}\mathbf{x}^T\mathbf{P}\mathbf{x}$, its derivative will be,

$$\dot{V} = \frac{1}{2}(\dot{\mathbf{x}}^T\mathbf{P}\mathbf{x} + \mathbf{x}^T\mathbf{P}\dot{\mathbf{x}})$$

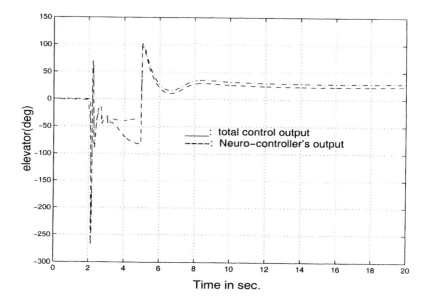

Figure 6.13. Controller's signal (NFC for model error 2)

$$\begin{aligned}
&= \frac{1}{2}(x^T A^T P x + x^T P A x) + \zeta x^T P g(x) \\
&= -x^T Q x + \zeta x^T P g(x) \\
&\leq -\|x\|[\lambda_{min}(Q)\|x\| - \zeta\|P\|\|g(x)\|]
\end{aligned} \quad (6.2)$$

Since,

$$lim_{\|x\|\to 0}\left(\frac{\|g(x)\|}{\|x\|}\right) = 0 \quad (6.3)$$

Given $p = \frac{\lambda_{min}(Q)}{\alpha\|P\|}$, $\exists \delta > 0$, when $\|x\| < \delta$, $\frac{\|g(x)\|}{\|x\|} < p$ stands, therefore the system is stable when the initial state x satisfies $\|x\| < \delta$. However, with the increase of ζ, the stable area will be smaller and the original x_0 will cause the instability of the overall system.

To expand the stability region, the proposed neuro-controller is utilized with the derived tuning rule. Fig.6.12 presents the comparison results of using different strategies: traditional weights tuning method *vs.* fully tuned strategy, the number of hidden neurons fixed *a priori vs.* GRBFN, etc. Fig.6.12(a) shows the tracking results. With only tuning the weights, the results diverge, while with tuning all the parameters, the performance is much better. Fig.6.12(b) shows the neuron history and by using a GRBFN, only 4 hidden neurons are

Aircraft Flight Control Applications Using Direct Adaptive NFC 111

recruited, which are less than using the normal RBFN. While in Fig.6.13, the control signals generated by GRBFN and the total control inputs are plotted, indicating how the neuro-controller learns the system dynamics on-line.

Further, since a GAP RBFN had already shown its effectiveness in controlling the linearized aircraft model (Section 6.1), it is also investigated in this example. Unfortunately a divergence is again observed under model error 2, which may be caused by the unproper pruning the hidden units. It is demonstrated that a pruning strategy may not be suitable when the nonlinearity is present in the system for this direct adaptive control architecture.

2.3. Discussion

It can be seen from this example that by learning the inverse dynamics of the system, the proposed NFC control scheme can tolerate more severe local nonlinearities than the conventional controller alone. It is worth noting that under model error 2, the conventional controller itself can not guarantee the stability of the overall system. This fact indicates that Assumption 2 in the derivation is actually a sufficient condition, but may not be a necessary one. However, further simulation results show that if ζ continuously increases, the RBFN controller can not learn the inverse dynamics of the system and a divergence is observed in the performance.

3. NFC for Full-Fledged Nonlinear 6-DOF Aircraft Model

In this section, the performance of the proposed neuro-flight-control (NFC) scheme is evaluated for flight controller design based on a high performance fighter aircraft executing a stability-axis roll maneuver at high angle-of-attack [56]. The stability-axis roll maneuver, which is completed by restoring the angle of attack to the original trim condition, consists of a vertical pull-up to increase and maintain a large angle of attack, followed by a rapid stability axis roll of $180°$. This maneuver is similar to the Herbst maneuver, although to a lesser extent due to the loss of control moments in deep stall conditions and the absence of thrust vector control in the aircraft under study. It is difficult to control such a maneuver because the normal acceleration tends to draw the heavy nose and tail portions of the aircraft farther from the axis of rotation at high roll rates, resulting in a positive pitch rate and an increasing α. The importance in designing a neural controller to control this maneuver lies in the fact that it does exercise the aircraft to a very wide dynamic range in a short time and brings out all the nonlinearities of the aircraft.

For validating the proposed neural controller and to assess the performance of the aircraft under large amplitude maneuvers, a full-fledged 6-DOF nonlinear aircraft model's flight simulation package has been developed on a PC using SIMULINK under the MATLAB environment [127]. For high fidelity, original wind tunnel test data is used and the aircraft dynamics is simulated in its original nonlinear form. The performance capabilities of the designed controller for executing the high α stability-axis roll maneuver is presented based on the results from this simulation package.

3.1. Full-Fledged Nonlinear 6-DOF Aircraft Model

The candidate aircraft is a high performance fighter aircraft similar to the F16 as reported in [87]. The aircraft model is built using Matlab/Simulink package [127] based on the wind tunnel data obtained from [87], which has the same dynamics as in Eq.(5.1). The mathematical symbols used to describe the aircraft dynamics is given as,

u, v, w	airspeed in the X, Y, Z axes of the body frame respectively;
p, q, r	roll, pitch, yaw rates about the X, Y, Z axes respectively;
ϕ, θ, ψ	roll, pitch, yaw angles about the X, Y, Z axes respectively;
I_x, I_y, I_z	moments of inertia about the X, Y, Z axes respectively;
I_{xz}	cross moment of inertia;
$C_{X,t}, C_{Y,t}, C_{Z,t}$	total aerodynamic force coefficients in X, Y, Z axes respectively;
$C_{l,t}, C_{m,t}, C_{n,t}$	total aerodynamic moment coefficients in X, Y, Z axes respectively;
$\bar{q}, \bar{c}$	dynamic pressure and mean chord respectively;
S, b, H_e	area of the wing; wing span; engine angular momentum;
α, β	angle-of-attack; sideslip angle;
μ	stability-axis roll rate;
V_t	total air speed;
a_n, a_y	normal and sideway acceleration respectively;
$\delta_h, \delta_a, \delta_r$	deflection angle of elevator, aileron, rudder respectively;
$\delta_{sb}, \delta_{lef}$	deflection angle of speed brake and leading edge flap;

Under the rigid body assumption and referenced to a body-fixed axis system, the mathematical equations describing the motions of the aircraft are as follows [117][86].

Force equations:
$$\dot{u} = rv - qw - g\sin\theta + \frac{\bar{q}S}{m}C_{X,t} + \frac{T}{m} \quad (6.4)$$

$$\dot{v} = pw - ru + g\cos\theta\sin\phi + \frac{\bar{q}S}{m}C_{Y,t} \quad (6.5)$$

$$\dot{w} = qu - pv + g\cos\theta\cos\phi + \frac{\bar{q}S}{m}C_{Z,t} \quad (6.6)$$

Moment equations:
$$\dot{p} = \frac{I_x - I_z}{I_x}qr + \frac{I_{xz}}{I_x}(\dot{r} + pq) + \frac{\bar{q}Sb}{I_x}C_{l,t} + \frac{H_e q}{I_x} \quad (6.7)$$

$$\dot{q} = \frac{I_z - I_x}{I_y}pr + \frac{I_{xz}}{I_y}(r^2 - p^2) + \frac{\bar{q}S\bar{c}}{I_y}C_{m,t} - \frac{H_e r}{I_y} \quad (6.8)$$

$$\dot{r} = \frac{I_x - I_y}{I_z}pq + \frac{I_{xz}}{I_z}(\dot{p} - qr) + \frac{\bar{q}Sb}{I_z}C_{n,t} + \frac{H_e q}{I_z}. \quad (6.9)$$

Aircraft Flight Control Applications Using Direct Adaptive NFC

Auxiliary equations:
$$\alpha = \tan^{-1} \frac{w}{u} \quad (6.10)$$
$$\beta = \sin^{-1} \frac{w}{V_t} \quad (6.11)$$
$$\dot{\mu} = p\cos\alpha + r\sin\alpha \quad (6.12)$$

These equations are highly nonlinear and coupled. The total aerodynamic coefficients $C_{X,t}$, $C_{Y,t}$, $C_{Z,t}$, $C_{l,t}$, $C_{m,t}$ and $C_{n,t}$ used in the simulation were derived from low-speed static and dynamic wind tunnel tests as in [87]. Total coefficient equations were used to sum the various aerodynamic contributions to a particular force or moment. For example, the X axis moment coefficient:

$$\begin{aligned} C_{l,t} = &\, C_l(\alpha,\beta,\delta_h) + \Delta C_{l,lef}(1 - \frac{\delta_{lef}}{25}) \\ &+ [\Delta C_{l,\delta_a} + \Delta C_{l,\delta_a,lef}(1 - \frac{\delta_{lef}}{25})]\frac{\delta_a}{20} + \Delta C_{l,\delta_r}\frac{\delta_r}{30} \\ &+ \frac{b}{2V_t}\{[C_{l_r}(\alpha) + \Delta C_{l_r,lef}(\alpha)(1 - \frac{\delta_{lef}}{25})]r \\ &+ [C_{l_p}(\alpha) + \Delta C_{l_p,lef}(\alpha)(1 - \frac{\delta_{lef}}{25})]p\} + \Delta C_{l_\beta}(\alpha)\beta \,. \quad (6.13) \end{aligned}$$

Other coefficients are given in similar forms (refer to [87]).

3.2. Kinematic and Navigation Equations

The equations of motion in Section 3.1 are based on body-fixed frame. In practical flight, the instantaneous attitude of the aircraft with respect to the ground-based coordinate is also important. This is usually described by *Kinematic Equations* and *Navigation Equations*.

Kinematic Equations

$$\dot{\phi} = p + \tan\theta \cdot (q\cdot\sin\phi + \dot{r}\cdot\cos\theta) \quad (6.14)$$
$$\dot{\theta} = q\cdot\cos\phi - r\cdot\sin\phi \quad (6.15)$$
$$\dot{\psi} = \frac{q\cdot\sin\phi + r\cdot\cos\phi}{\cos\theta} \quad (6.16)$$

where ϕ, θ and ψ are the three Euler angles, *i.e.*, roll, pitch and yaw angles. These angles describe the relative angular position between body-fixed frame and flat-earth reference frame. The disadvantage of the above three Euler equations is that, when $\cos\theta$ is zero, *i.e.* $\theta = \pm 90°$, $\dot{\psi}$ is infinite which causes numerical difficulties in computer simulation. So the above Euler equations are only used in the trimming and linearization programs at flight conditions where θ are not equal to or near $\pm 90°$. To avoid this difficulty, another method, the so-called quaternion four-variable representation, is adopted for the simulation program introduced later.

Quaternion Equations

$$\begin{bmatrix} \dot{q}_0 \\ \dot{q}_1 \\ \dot{q}_2 \\ \dot{q}_3 \end{bmatrix} = \frac{1}{2} \begin{bmatrix} 0 & p & q & r \\ -p & 0 & -r & q \\ -q & r & 0 & -p \\ -r & -q & p & 0 \end{bmatrix} \begin{bmatrix} q_0 \\ q_1 \\ q_2 \\ q_3 \end{bmatrix} \quad (6.17)$$

where

$$q_0 = \pm(\cos(\frac{\phi}{2})\cos(\frac{\theta}{2})\cos(\frac{\psi}{2}) + \sin(\frac{\phi}{2})\sin(\frac{\theta}{2})\sin(\frac{\psi}{2})) \quad (6.18)$$

$$q_1 = \pm(\sin(\frac{\phi}{2})\cos(\frac{\theta}{2})\cos(\frac{\psi}{2}) - \cos(\frac{\phi}{2})\sin(\frac{\theta}{2})\sin(\frac{\psi}{2})) \quad (6.19)$$

$$q_2 = \pm(\cos(\frac{\phi}{2})\sin(\frac{\theta}{2})\cos(\frac{\psi}{2}) + \sin(\frac{\phi}{2})\cos(\frac{\theta}{2})\sin(\frac{\psi}{2})) \quad (6.20)$$

$$q_3 = \pm(\cos(\frac{\phi}{2})\cos(\frac{\theta}{2})\sin(\frac{\psi}{2}) - \sin(\frac{\phi}{2})\sin(\frac{\theta}{2})\cos(\frac{\psi}{2})) \quad (6.21)$$

Whichever sign is chosen in the above equations, the same choice must be used for all. So the Euler angles are given by

$$\phi = \mathrm{atan2}(b_{23}, b_{33}) \quad (6.22)$$
$$\theta = -\sin^{-1}(b_{13}) \quad (6.23)$$
$$\psi = \mathrm{atan2}(b_{12}, b_{11}) \quad (6.24)$$

where b_{ij} is the ith row and jth column of the following matrix B and atan2 is the MatLab four-quadrant inverse function for $\tan(\cdot)$ function.

$$B = \begin{bmatrix} q_0^2 + q_1^2 - q_2^2 - q_3^2 & 2(q_1 q_2 + q_0 q_3) & 2(q_1 q_3 - q_0 q_2) \\ 2(q_1 q_2 - q_0 q_3) & q_0^2 - q_1^2 + q_2^2 - q_3^2 & 2(q_2 q_3 + q_0 q_1) \\ 2(q_1 q_3 + q_0 q_2) & 2(q_2 q_3 - q_0 q_1) & q_0^2 - q_1^2 - q_2^2 + q_3^2 \end{bmatrix} \quad (6.25)$$

The initial values of B matrix elements can be set up by using initial values of Euler angles. The nine elements of the B matrix can also be given in terms of the Euler angles.

$$b_{11} = \cos\theta\cos\psi \quad (6.26)$$
$$b_{21} = \sin\phi\sin\theta\cos\psi - \cos\phi\sin\psi \quad (6.27)$$
$$b_{31} = \cos\phi\sin\theta\cos\psi + \sin\phi\sin\psi \quad (6.28)$$
$$b_{12} = \cos\theta\sin\psi \quad (6.29)$$
$$b_{22} = \sin\phi\sin\theta\sin\psi + \cos\phi\cos\psi \quad (6.30)$$

$$b_{32} = \cos\phi\sin\theta\sin\psi - \sin\phi\cos\psi \quad (6.31)$$
$$b_{13} = -\sin\theta \quad (6.32)$$
$$b_{23} = \sin\phi\cos\theta \quad (6.33)$$
$$b_{33} = \cos\phi\cos\theta \quad (6.34)$$

which are very useful in calculations of the following Navigation Equations.
Navigation Equations

$$\dot{P}_N = u \cdot \cos\theta\cos\psi + v \cdot (-\cos\phi\sin\psi + \sin\phi\sin\theta\cos\psi) +$$
$$w \cdot (\sin\phi\sin\psi + \cos\phi\sin\theta\cos\psi) \quad (6.35)$$
$$\dot{P}_E = u \cdot \cos\theta\sin\psi + v \cdot (\cos\phi\cos\psi + \sin\phi\sin\theta\sin\psi) +$$
$$w \cdot (-\sin\phi\cos\psi + \cos\phi\sin\theta\sin\psi) \quad (6.36)$$
$$\dot{h} = u \cdot \sin\theta - v \cdot \sin\phi\cos\theta - w \cdot \cos\phi\cos\theta \quad (6.37)$$

With the quaternion expression, the navigation equations can be written as follows

$$\dot{P}_N = u \cdot b_{11} + v \cdot b_{21} + w \cdot b_{31} \quad (6.38)$$
$$\dot{P}_E = u \cdot b_{12} + v \cdot b_{22} + w \cdot b_{32} \quad (6.39)$$
$$\dot{h} = -(u \cdot b_{13} + v \cdot b_{23} + w \cdot b_{33}) \quad (6.40)$$

The three variables (h, P_E, P_N) are for ground-based system under the flat-earth assumption with P_E, P_N and h begin the aircraft position in east, north and height, respectively. These three variables can be used in describing the aircraft flight path of the simulation results.

3.3. Auxiliary Equations

In addition to the equations given so far, a few other important physical quantities: angle of attack (α), sideslip angle (β), resultant velocity (V_t) and normal acceleration (a_n) are given by the following auxiliary equations.

$$V_t = \sqrt{u^2 + v^2 + w^2} \quad (6.41)$$
$$\alpha = \tan^{-1}(\frac{w}{u}) \quad (6.42)$$
$$\beta = \sin^{-1}(\frac{v}{V_t}) \quad (6.43)$$
$$a_n = \frac{qu - pv + g\cos\theta\cos\phi - \dot{w}}{g} \quad (6.44)$$

Please refer to Stevens and Frank Lewis's book [117] and Nelson's book [86] for detailed derivations of all the above equations.

3.4. Other Equations

Stability roll and yaw rate

Future high performance aircraft must have the ability to maneuver rapidly at high angular roll rate (μ_{rat}) under high angle of attack (α) conditions. This kind of roll can lead to some highly undesirable effects which can be described as *Kinematic coupling* and *Inertial coupling*.

Consider the effect of rapid 90° body-axis roll under a high angle of attack (α) flight. It is easy to visualize that the angle of attack will be converted immediately, and almost entirely, into a sideslip angle (β). This is referred to as *Kinematic coupling* of alpha (α) to beta (β) transition which is undesirable in practical flight.

The sideslip caused by kinematic coupling is referred to as adverse sideslip, and will tend to oppose the roll. This is because that a right roll will generate a positive β through kinematic coupling and hence will lead to a negative rolling moment which is generated by vertical tail. The sideslip will exist until the aircraft is rolled into the original wind axis once more. A large sideslip angle is undesirable for several reasons.

1 The effectiveness of aerodynamic control surfaces may be greatly reduced;

2 Directional stability may be lost, so that in some cases, the aircraft will tumble from end to end.

3 A large sideforce may be generated which will lead to overload on pilot, and may possibly bread the vertical tail of the aircraft.

To avoid such coupling, the aircraft must be able to perform the roll motion around the stability axis with sideslip angle (β) nearly zero under such a high angle of attack (α) flight condition. This flight also faces another coupling dynamics. Since in future fighter aircraft design, the mass of the aircraft will be concentrated in the fuselage to get high performance in rapid rolling flight, the aircraft can be considered as a big "dumbbell". If the aircraft performs a stability roll flight, no matter it is a positive one or negative one, a pitch up moment is generated which will force the angle of attack (α) to diverge. This is the so called Inertial coupling which, together with the above Kinematic coupling problem, must be considered seriously in controller design.

For the considerations of the above two coupling problems, in modern fighter aircraft design, the control system must provide the ability to roll the aircraft about the stability axis with the initial α maintained and β nearly zero. This can be explained as a tracking problem, which means that the aircraft will track stability-axis roll rate P_{stab} and yaw rate r_{stab}.

$$P_{stab} = p \cdot \cos\alpha + r \cdot \sin\alpha \qquad (6.45)$$

$$r_{stab} = -p \cdot \sin\alpha + r \cdot \cos\alpha \qquad (6.46)$$

Aircraft Flight Control Applications Using Direct Adaptive NFC

where p_{stab} is the desired roll rate and r_{stab} should be kept as small as possible. In order to distinguish between stability roll rate p_{stab} and air static pressure P_s, the stability roll rate is usually denoted by μ_{rat}.

Fig.6.14 shows the full-fledged nonlinear high performance aircraft model constructed by using block diagrams with Simulink in the Matlab environment [127]. Because the calculations of all the aerodynamic coefficients ($C_{x,t}$, $C_{y,t}$, etc.) and engine thrust are based on the experiment data which are listed in tabular forms, a lookup-table program is included.

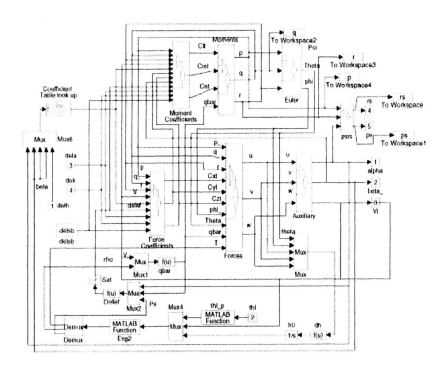

Figure 6.14. The high performance fighter aircraft model block

This model contains all the dynamic equations (excluding the engine model and the aerodynamic coefficient lookup-table) of aircraft as in Eq.(6.4) to Eq.(6.9). It is further subdivied into the force block, the moment block, the aux block, etc., for detail refer to [127]. The aerodynamic moment coefficients are referenced to the nominal center of gravity location of $X_{cg} = 0.35\bar{c}$ and are corrected to the desired center-of-gravity position in the coefficient equa-

118 FULLY TUNED RBF NEURAL NETWORK FOR FLIGHT CONTROL

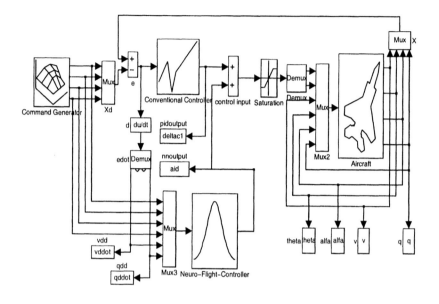

Figure 6.15. Block diagram for the Neuro-Flight-Controller

tions. The aircraft is powered by an afterburning turbofan jet engine. A typical model for the engine simulation is given by describing the thrust responses to throttle inputs. The thrust values are given in tabular form for idle, military and maximum thrust values. Moreover, the dynamics of the actuator that drives the control surfaces is modeled as simple first-order transfer function. For details refer to [87].

For the candidate fighter aircraft used in this study, there is no provision of thrust vectoring of control and hence the maneuver has to be restricted to within the capabilities of the conventional control surfaces. Since the leading edge flaps δ_{lef} are designed as auto-scheduled according to α, hence the available control surfaces are δ_h, δ_a and δ_r. Also, in a high α stability-axis roll maneuver, throttle is usually pushed to nearly full position to prevent significant loss of speed, so the engine thrust control is not considered in the controller design. As can be seen from the aircraft dynamics, the mathematical equations describing the motions of the aircraft are the same as Eq.(5.1), where the state vector $\mathbf{x} = [u, v, w, p, q, r, \alpha, \beta, \mu]^T$ with $\mathbf{x_t} = [\alpha, \beta, \mu]^T$, and $\mathbf{u} = [\delta_h, \delta_a, \delta_r]^T$. The complete aerodynamic data acquired from the wind tunnel test covers a wide range of $-20° < \alpha < 90°$, $-30° < \beta < 30°$.

Aircraft Flight Control Applications Using Direct Adaptive NFC 119

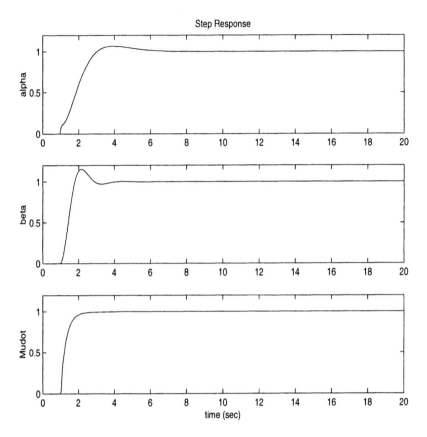

Figure 6.16. Step responses of the conventional controller based on the linearized aircraft model (units of α, β in degree, and $\dot{\mu}$ in degree per second)

3.5. High α Stability-Axis Roll Maneuver

The high α stability-axis roll (Herbst-like) maneuver starts with straight level flight at the trim condition of $\alpha = 2.37°, V_t = 152.4 m/s, h = 1000 m, \delta_h = -0.5803°$ (position A). A pitch-up command to the elevator is given to increase the α from its trim value to $30°$ at $3s$. The command for $\dot{\mu}$ starts at $5s$ and consists of three regions: the rising region (approximately $1s$), the holding region (approximately $3s$) and the arrest region (approximately $1s$). The stability axis roll rotates the aircraft about its stability axis so as to turn the direction of flight. Then α is decreased to bring the nose down to the initial α at $18s$. This maneuver is similar to the Herbst maneuver [32], although to a lesser extent due to the loss of control moments in deep stall conditions and the absence of thrust

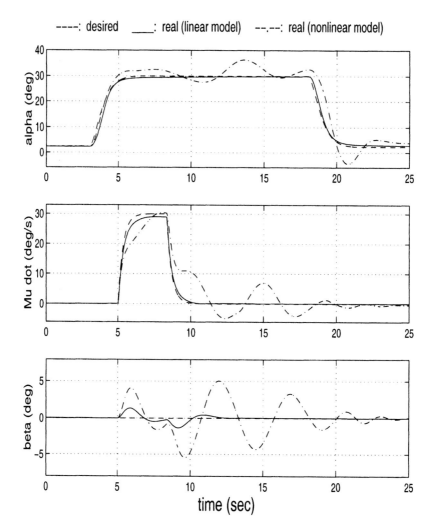

Figure 6.17. Performance of the proportional controller for nonlinear aircraft model

vector control in the aircraft under study. This maneuver is difficult to control because the normal acceleration tends to draw the heavy nose and tail portions of the aircraft further from the axis of rotation at high roll rates, resulting in a positive pitch rate and an increasing α.

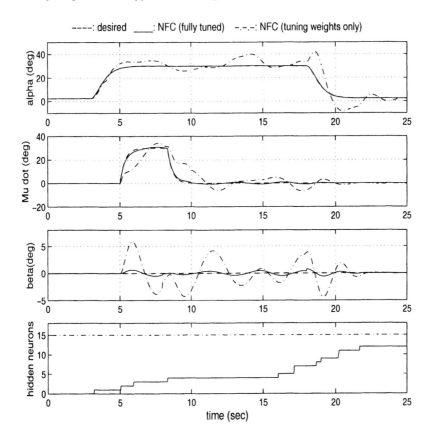

Figure 6.18. Performance of the NFC for nonlinear aircraft model

3.6. Simulation Studies

Fig.6.15 illustrates the SIMULINK model that is built for evaluating the proposed NFC control strategy based on the full-fledged high performance fighter aircraft model.

The figure represents the root layer model. It can be seen that the model contains 4 main blocks, namely, command generator, conventional controller, neuro-flight-controller and the aircraft block. Since the software package of the full-fledged aircraft model shown in Fig.6.14 has been validated to possess high fidelity in simulating the open loop and close loop nonlinear aircraft dynamics, simulation studies for performing the high α stability-axis roll maneuver using the designed NFC is carried out based on this package and it is embedded into the aircraft block. The command generator is built to produce the desired tra-

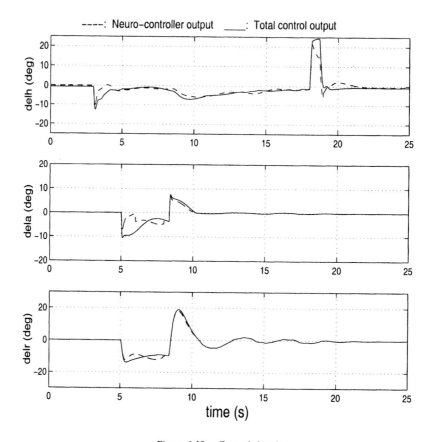

Figure 6.19. Control signals

jectories and in this case the desired α, μ, β signals are generated. Moreover, the conventional controller and the neuro-flight-controller designed are incorporated into the conventional controller block and the NFC block, respectively.

In this study, first a conventional proportional controller is designed as a baseline controller to achieve desired response characteristics for the aircraft linearized at a chosen trim point. For this purpose trimming and linearizing of the nonlinear model is carried out to obtain linear models suitable for conventional controller design. To satisfy the Assumption 2, a high α, low speed trim condition is selected as the design point for this conventional controller: $V_t = 70m/s, \alpha = 15.6474°, h = 1000m, \delta_h = 0.5851°$. We name this trim point as A.

Aircraft Flight Control Applications Using Direct Adaptive NFC 123

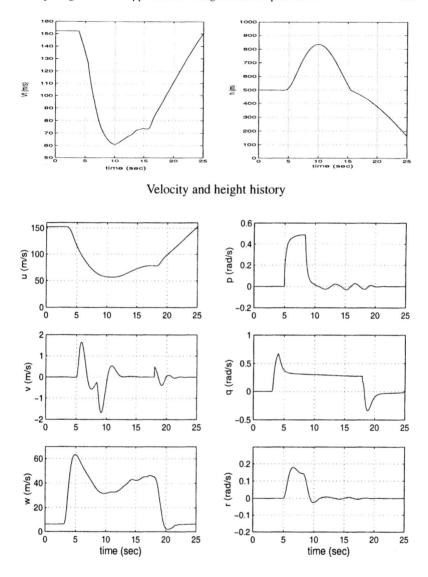

Velocity and height history

Figure 6.20. State variables time history

Before proceeding further, it is instructive to evaluate the performance of the conventional controller alone. The conventional controller is designed to provide good control performance with the linearized aircraft model at the point

124   FULLY TUNED RBF NEURAL NETWORK FOR FLIGHT CONTROL

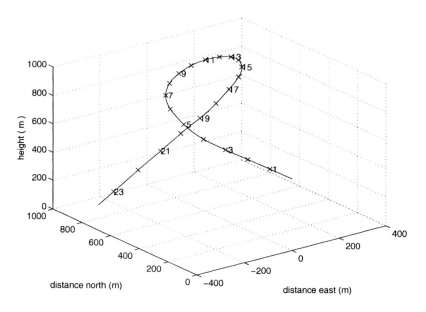

Figure 6.21. Actual flight profile in three dimension (fully tuned GRBFN)

A. The step response of the conventional controller with the linearized aircraft model at point A is shown in Fig.6.16, and the response of the conventional controller to the full-fledged 6-DOF aircraft model is presented in Fig.6.17. It can be seen from the figure that although this controller performs quite well for the linear model, the tracking performance deteriorates greatly when applied to the 6-DOF nonlinear model (dash-dot line) and the deviation in β is high (5°).

Hence, the RBFN controller is added in parallel to learn the inverse dynamics of the system and increase the tracking accuracy. The proposed feedback-error-learning strategy and the derived stable tuning rule are utilized to implement this Herbst-like maneuver. Fig.6.18 presents the results of using the proposed neuro-flight-controller. A comparison of using the (proposed) full tuning strategy (GRBF) with that of tuning only the weights of the RBFN is also presented. For the RBFN controller with only weights tuning, the number of hidden neurons are fixed as 15 prior according to experience and the centers of the hidden units are determined as 0.25 according to a grid method described in (Fabri, & Kadirkamanathan, 1996). From Fig.6.18 it is obvious that the performance is poor for the case of only tuning the weights (dash-dot line). However, by tuning all the parameters of the GRBFN, good tracking performance is achieved for both of the α and μ with small variation in β (less than 1° as required) during the maneuver (line). It is also noted that except for the better accuracy, less hidden neurons is recruited using the GRBFN – 12 as compared with the 15 by using the normal RBFN. Fig.6.19 displays the control signals, the continuous line representing the total control signals and the dashed line representing the neuro-controller's outputs. It can be seen from the figure that the control inputs generated are well within the actuator limits without saturation ($\delta_h(max) = \pm 25°, \delta_a(max) = \pm 20°, \delta_r(max) = \pm 30°$). Also it can be seen how the neural control outputs change when the commands are executed indicating the learning process. Fig.6.20 plots the total velocity and altitude at the top and other aircraft states such as u, v, w, p, q, r are also shown. These states are back to the trimmed condition at the end of the maneuver. Fig.6.21 presents the trajectory of the maneuver in 3 dimensions using the neuro controller with the fully tuned RBF neural network. The numbers on the trajectory line indicate the time sequence.

In summary, a stable on-line learning control strategy using GRBFN to execute a high α stability-axis roll maneuver has been studied. Simulation studies based on a high performance fighter aircraft model demonstrate that with the developed control strategy and the proposed tuning rule, a more compact neural network can be implemented with better performance.

4. Summary

Based on the three simulation studies presented in this chapter, it can be concluded that the proposed direct adaptive control strategy incorporating a fully-tuned RBFN controller is more efficient than the traditional feedback-error-learning scheme with only tuning the weights. Moreover, simulation results also demonstrated that with a growing strategy, a more compact network structure can be implemented compared with the network with fixed number of hidden neurons.

Chapter 7

MRAN NEURO-FLIGHT-CONTROLLER FOR ROBUST AIRCRAFT CONTROL

In the previous chapters, the direct adaptive control strategy with the tuning rule derived from the Lyapunov stability theory was presented. In this chapter, the recently proposed MRAN algorithm [61] using a fully tuned RBFN is investigated to control a linearized aircraft model. It is well known that with a pruning strategy and EKF tuning rules, MRAN can implement a more compact network structure. This superiority has been demonstrated in many applications such as function approximation, pattern classification and nonlinear system identification. However, it is the first time an attempt to explore the use of MRAN algorithm for aircraft control applications is carried out [57].

1. Problem Formulation and Conventional Controller

A linearized longitudinal dynamic model of the F8 aircraft, which is represented by Eq.(4.1) and discussed by Sadhukhan *et al.* [105], is utilized for studying the performance of the MRAN controller in this chapter.

In the previous chapters, by incorporating a fully tuned RBFN controller, we investigated the performance of the proposed indirect and direct adaptive control strategies for the linearized F8 aircraft model, respectively. The objective of the control is to track the desired velocity and pitch rate. In the indirect adaptive control (Section 5.2.3), the study is carried out based on the nominal model of the aircraft model. In the direct adaptive control case using the Lyapunov-tuning rule (Section 7.1), a robust controller is evaluated when the aircraft is undergoing model error. Further in this chapter, the aim is to design a robust MRAN controller which has fault tolerant characteristics, that is, with the controller designed, the aircraft can follow the desired trajectories not only in the nominal condition, but also when undergoing large model errors or sudden actuator sluggishness.

In [105], studies were carried out to evaluate the response in the presence of model error and actuator sluggishness which was of lower levels. For example, under 40% of the model error, the velocity oscillated in the first few seconds though settled eventually. While for actuator sluggishness, the controller can tolerate a sudden change in the elevator time constant from 0.1 to 4s. However, in this study, with the traditional feedback controller and the RBF controller's on-line learning ability (MRAN algorithm), it is shown that the proposed scheme can tolerate more serious fault conditions. To assess the performance of the MRAN flight-controller, different flight conditions and faults have been simulated. Results for the following two scenarios are presented.

- **Scenario 1** Aircraft modeling error: This is simulated by changing all the elements of the F_1 and G_1 by 50% and 70% respectively, for $t \geq 0s$

- **Scenario 2** Elevator actuator sluggishness: This has been simulated by suddenly changing the elevator time constant from 0.1 to 8s, from time $t \geq 4s$.

Using conventional controller design methods, a proportional controller is designed first as a baseline controller to provide stability augmentation. However, under the first failure scenario (model error of 70%), velocity response is very slow (Fig.7.1(a)), pitch rate response is bad as can be seen in Fig.7.1(b), where large steady state error exists during the maneuver. Under scenario 2 (sudden variation of elevator actuator time constant), the velocity still follows the desired trajectory quite well, however, the pitch rate response after 4 seconds can not catch up with the desired signal.

Thus it can be clearly seen that the conventional controller is not able to perform well under the two fault scenarios indicated. Of course, on-line tuning of the proportional controller's parameters may yield better results, but it is a time-consuming work which is not practical in aircraft application. Under this situation, neural control offers an attractive alternative.

2. Robust MRAN-Flight-Controller

Since the conventional controller designed under nominal condition can not cope with the problem of model error or actuator sluggishness, the scheme of MRAN aided conventional controller as shown in Fig.7.2 has been developed. Simulation results under the above two fault scenarios with the proposed MRAN neuro-flight-controller are presented. To demonstrate the advantage of using MRAN algorithm, a meaningful comparison (tracking accuracy, resulting network size) between MRAN neuro-flight-controller using GAP Grownig and Pruning (GAP) RBF network and RAN-EKF using GRBFN neuro-controller is also presented.

MRAN Neuro-Flight-Controller For Robust Aircraft Control

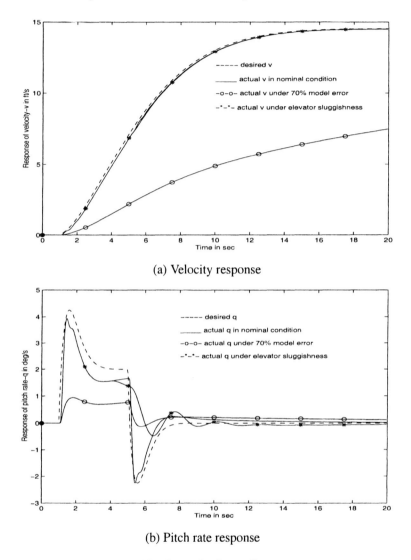

(a) Velocity response

(b) Pitch rate response

Figure 7.1. Conventional controller response

In this architecture, MRAN's on-line learning is also based on the well-known "feedback error learning" scheme. The scheme uses a conventional proportional controller in the inner loop to stabilize the system dynamics, and the MRAN neuro-flight-controller acts as an aid to the conventional controller

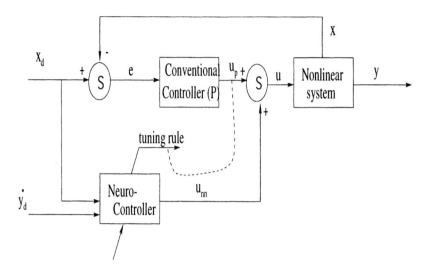

Figure 7.2. Control structure of MRAN controller

through on-line learning. Thus the total control signal to the aircraft is a sum of proportional and MRAN control signals. In addition, to satisfy the physical limitations of the actuators, a limiter is also included in the system. This control architecture is similar to the one proposed in Chapter 6, however, in using the MRAN controller, the tuning rule is based on the extended Kalman filter (EKF), and to put it simply, the error for updating the MRAN network is based on the conventional controller output signal as it truly reflects the error between the commands and actual outputs [27]. The MRAN network starts from a zero neuron network and builds up the structure based on this error. At the initial period, the conventional controller signal is the main signal controlling the system. When a faulty condition is encountered, MRAN controller learns the needed control inputs quickly through on-line learning and generates a larger control signal, driving the aircraft to follow the desired outputs. Actually, in this process the MRAN neuro-flight-controller itself replaces the proportional controller.

3. Simulation Results

To facilitate a quantitative analysis, an average error E_{ave} is defined as a criterion to assess the tracking accuracy,

$$E_{ave} = \sum_{i=1}^{N} \frac{\sqrt{|E_v|^2 + |E_q|^2}}{N} \quad (7.1)$$

where $N = 1000$ is the number of samples along the time history. E_v, E_q are the tracking error for velocity and pitch rate, respectively, and the unit for E_v is ft/sec, E_q is deg/s.

3.1. Response to Model Error

With the RAN-EKF and MRAN neuro-flight-controller, the results for tracking desired signals under 70% model error are presented in Fig.7.3.

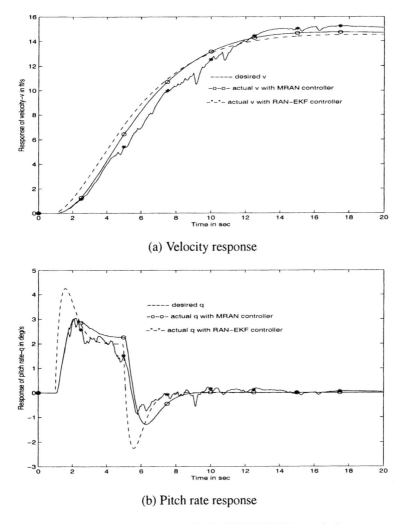

(a) Velocity response

(b) Pitch rate response

Figure 7.3. Tracking result of RAN-EKF & MRAN (scenario 1)

132 FULLY TUNED RBF NEURAL NETWORK FOR FLIGHT CONTROL

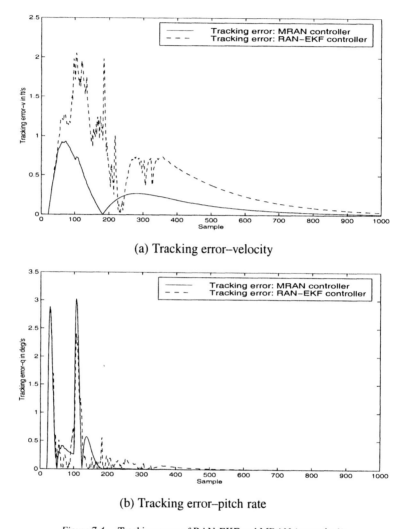

(a) Tracking error–velocity

(b) Tracking error–pitch rate

Figure 7.4. Tracking error of RAN-EKF and MRAN (scenario 1)

From the figure it is clear that with the MRAN neuro-flight-controller, the outputs of the velocity and pitch rate follow the desired commands closely. However, with the MRAN algorithm, the neuro-controller achieves better tracking accuracy, which is reflected in Fig.7.4, where the tracking error of velocity and pitch rate using the MRAN and RAN-EKF controller is compared.

Fig.7.5 shows the evolution of hidden neurons. After reaching the steady state, RAN-EKF uses 5 hidden neurons, while with a pruning strategy, MRAN

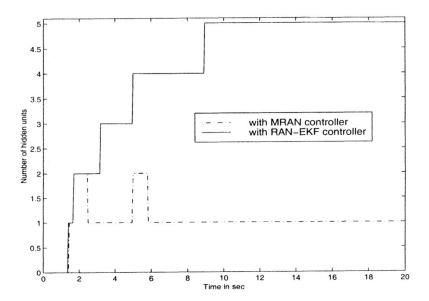

Figure 7.5. Hidden neuron Comparison using RAN-EKF and MRAN (scenario 1)

uses only 1 hidden neuron. The results indicate that using MRAN algorithm, the neuro-controller can implement a more compact network structure, resulting in fast on-line learning. A quantitative comparison of the performance is given in Table 7.1, which also includes the comparison results of the two neuro controllers under 50% model error. From the table it can be seen that for 50% model error, with RAN-EKF neuro-controller, the average error E_{ave} is 0.16, while using MRAN, only 1 hidden neuron is needed with the average error $E_{ave} = 0.14$.

Table 7.1. Performance comparison: MRAN and RAN-EKF neuro-controller

Scenario	Elevator sluggishness			Model error					
	0.1s -- -8s			50%			70%		
Algorithm	E_{ave}	H_m	H_l	E_{ave}	H_m	H_l	E_{ave}	H_m	H_l
RAN-EKF	.32	20	20	.16	4	4	.50	5	5
MRAN	.26	7	1	.14	2	1	.24	2	1

Note:
$H_m (H_{max})$ is the maximum number of hidden neurons used in the procedure
$H_l (H_{last})$ is the number of hidden neurons used when the maneuver finishes

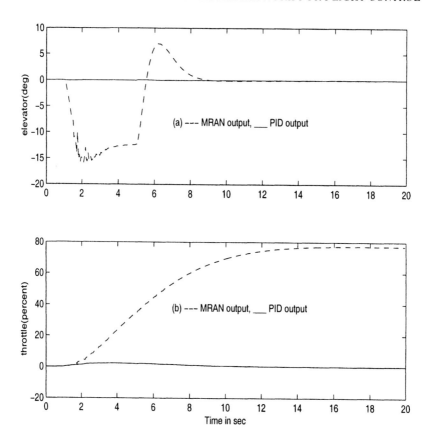

Figure 7.6. MRAN vs. conventional controller's output (scenario 1)

Fig.7.6 depicts the control signals for both the proportional controller and MRAN controller. From the figure it is obvious that MRAN produces a larger control signal than that of the proportional controller to achieve the accurate tracking results. It should also be noted that the conventional controller alone was not able to perform this function earlier (refer to Fig.7.3). It can also be seen that the MRAN control signals are within the physical limits and do not reach saturation.

Although simulation studies show that MRAN can also be applied to control the system even when model error is greater than 70%, the results indicate that elevator actuator saturation will be the limiting factor resulting in deterioration of the tracking performance.

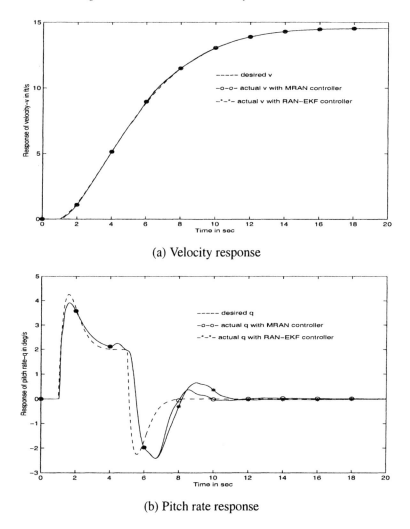

Figure 7.7. Tracking result of RAN-EKF and MRAN (scenario 2)

In Chapter 7, instead of using the MRAN neuro-flight-controller, a direct adaptive control strategy using the fully tuned GRBFN controller and derived Lyapunov based tuning rule is also investigated for control the linearized F8 aircraft under 70% model error. Compared with the results obtained in Chapter 7, it is noted that using the MRAN strategy and the EKF tuning rule, the performance is better, *i.e.*, MRAN only uses one hidden neuron at the end of

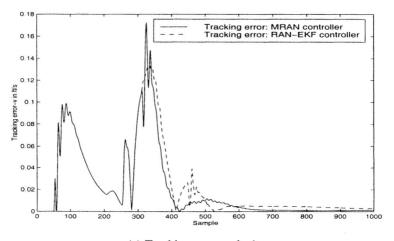

(a) Tracking error–velocity

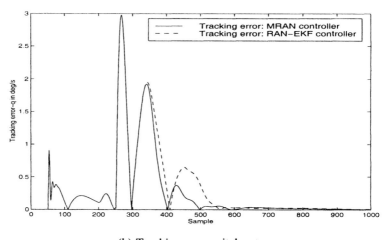

(b) Tracking error–pitch rate

Figure 7.8. Tracking error with RAN-EKF and MRAN (scenario 2)

the control, while using the Lyapunov based tuning rule, 4 hidden neurons are added. In addition, using the MRAN algorithm, the tracking error is also small.

3.2. Response to Actuator Sluggishness

In this study, a more severe actuator sluggishness condition is considered. The elevator time constant is changed from $0.1sec$ to $8sec$ suddenly after 4

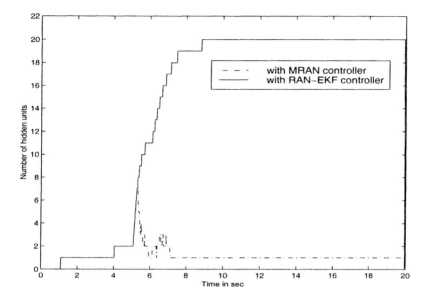

Figure 7.9. Comparison of hidden neuron using RAN-EKF and MRAN

seconds of the maneuver. RAN-EKF and MRAN controller are studied in this simulation, and their performances are compared.

Fig.7.7 gives the command responses for the velocity and pitch rate. From the figure it is evident that with the RAN-EKF or MRAN neuro-flight-controller, pitch rate can be followed with greatly improved accuracy. A detailed quantitative comparison between the RAN-EKF controller and MRAN controller is also shown in Table 7.1.

In this example, although it is difficult to see any difference in the tracking error with using MRAN and RAN-EKF algorithm, a more pronounced MRAN behavior is observed specifically in its ability to automatically delete those neurons that contributes little to the output, thus realizing a parsimonious network structure. From Fig.7.9 the RAN-EKF uses 20 hidden units, while for MRAN the maximum number of hidden units used is only 7 and after reaching the steady state, this number drops to 1.

It can also be seen that the MRAN controller adds neuron at two key points. One is at 4 second when the actuator sluggishness increase occurs. Since there is only a small kink to the system outputs, the network just adds several neurons. In the following several seconds, the aircraft first decreases the positive pitch rate rapidly until it reaches a negative value and then quickly returns to zero. The whole procedure only lasts for about 3 seconds. To fulfill this ma-

138   FULLY TUNED RBF NEURAL NETWORK FOR FLIGHT CONTROL

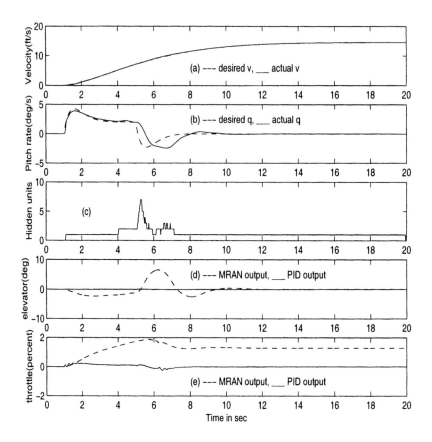

Figure 7.10. Performance analysis of MRAN neuro-flight-controller (scenario 2)

neuver, the elevator should swing up and down sharply. Because of the actuator sluggishness, with no neuro-controller, it can be seen from Fig.7.7 that after 4 seconds, the pitch rate can not be followed. While in Fig.7.7, with the RBF controller (using either RAN-EKF algorithm or MRAN algorithm), the pitch rate can be followed with a little delay. Note that the delay is caused because MRAN controller needs more time to grasp the changed system dynamics.

Fig.7.10 depicts the performance of using MRAN neuro-flight-controller. Fig.7.10(c) clearly shows the functioning of MRAN controller as it builds up the neurons at 4s and 5s when the failure occurs and quickly settles desired network size. Fig.7.10(d) and (e) show the conventional and MRAN control signals for both throttle and elevator, it can be seen that the MRAN controller performs an on-line learning of the changed system dynamics and tries to make the outputs follow the desired commands. It also should be noted that these signals have not reached saturation and are well within the physical limits.

Table 7.1 indicates that with a MRAN neuro-flight-controller, in all the scenarios, better performance than the RAN-EKF neuro-controller is observed (the average tracking error E_{ave} is smaller) with less hidden neurons, that is, both the number of the maximum neurons H_{max} added in the procedure and the number of hidden units used after reaching the steady state (H_{last}) is less. In this way, with the increasing/decreasing ability, MRAN neuro-flight-controller can realize a more compact network structure, which is suitable for the on-line implementation.

4. Summary

Based on the simulation studies, the following points on MRAN flight control strategy can be summarized.

1 The traditional controller plays an important role in the proposed strategy. It is not only used to stabilize the closed loop system under all cases, but also provide the error signals to train the network. With the traditional controller, the scheme is more robust.

2 As can be seen in fault Scenario 1 and 2, neuro-controller can not output a control signal with abrupt change, therefore the actual output lags behind the desired signal (especially in scenario 2, where the neuro-controller starts to learn the changed dynamics from 4 second). Based on this fact, it is recommended that a simple (level 2 or level 3 handling qualities requirements) desired dynamics filter be used during the first few seconds of flight.

3 Since the primary control target is to follow the desired velocity and pitch rate accurately, only these two variables are fed back to form the roughly designed proportional controller.

4 The comparison study of MRAN with RAN-EKF demonstrates that with a increasing or decreasing mechanism, MRAN is very suitable for real implementation of on-line control, and its performance can also be investigated with other control strategies.

5 The performance of MRAN algorithm is also better than that of the fully tuned RBFN controller using Lyapunov-based tuning rule, *i.e.* less hidden neuron and better tracking error.

It can be concluded from the simulation studies presented in this chapter that the proposed direct adaptive control strategy are demonstrated to be efficient than the traditional feedback-error-learning scheme that only tuning the weights of the network. Moreover, MRAN neuro-flight-controller can endure serious fault conditions with only on-line learning and no prior off-line training. Unfortunately, although simulation results show that the proposed scheme is better, there is no strict mathematical proof to guarantee the stability and performance for this scheme. Therefore, further research work should be carried out on this aspect.

Chapter 8

CONCLUSIONS AND FUTURE WORK

1. Conclusions

This book has described an in-depth investigation into designing adaptive controllers based on fully tuned RBF networks and their applications in the field of aircraft flight control. The conclusions from this study can be summarized as follows.

- A new stable on-line identification scheme using a fully tuned GRBF network has been developed for the identification of nonlinear systems with external inputs. To guarantee the stability of the overall system and the convergence of the identification errors, the parameter tuning rule for updating all the parameters of the RBFN has been derived using a Lyapunov synthesis approach. Compared to the traditional identification method where only the weights of the RBFN are adaptable, the proposed method avoids the prefixing of centers and widths of the Gaussian function, and hence approximates the system dynamics more accurately.

The performance of the proposed stable identification scheme has been evaluated through a comparative study. Based on the identification of a complicated nonlinear missile dynamics, it has been shown that the performance of the proposed scheme is better, *i.e.* smaller identification error and less hidden neurons can be obtained.

- In identification based indirect adaptive control, we have investigated the off-line training/on-line control strategy and the on-line learning/on-line control strategy. For evaluating the neuro-controller's performance, a linearized aircraft model was utilized in the simulation studies. Simulation study indicates that *a priori* off-line training is required in the indirect adaptive control case, and using the on-line learning/on-line control strategy, the performance is poor due to lack of prior knowledge obtained from off-line training.

For off-line training/on-line control scheme, it was demonstrated that the adding /pruning strategy adopted in off-line training can implement compact network structure, and the randomly arranged training data pair provides the best command following performance during on-line simulation. In addition, when the number of hidden neurons in the RBFN controller are fixed or grow automatically, good tracking results are observed. However, the performance deteriorates, or even diverges if a pruning strategy is incorporated.

- A new direct adaptive control scheme using the fully tuned RBFN has been developed for nonlinear adaptive control. The tuning rule for updating all the parameters of the RBFN is derived based on the Lyapunov stability theory, which guarantees the stability of the overall system. This RBFN based neuro-controller extends Kawato's conventional feedback-error-learning scheme where only the weights of the network are adaptable, and outperforms Kawato's method with better tracking accuracy. A robustness analysis demonstrates that the proposed neuro-controller can tolerate large approximation errors and model errors. Since no off-line training is required in this new adaptive control scheme and the neuro-controller designed can compensate the system nonlinearities by on-line learning, this original contribution made in this book can be applied in a number of real applications, *i.e.* robot control, aircraft flight control, etc..

- This book has concentrated mainly on the applications of the proposed adaptive neuro-controller in the field of aircraft flight control. Simulation studies are carried out based on three different control objectives and aircraft models, which includes command following for a linearized F8 aircraft model in longitudinal mode (linearized model), pitch-rate control for a localized nonlinear fighter aircraft model (localized nonlinear model), and implementing a high α stability-axis roll maneuver based on a full-fledged 6-DOF nonlinear high performance aircraft model (full nonlinear model). The simulation results indicate the effectiveness of the developed direct adaptive architecture and the derived adjusting law. It is concluded from the results that using the fully tuned RBFN controller in a direct adaptive control scheme, the tracking accuracy improves greatly.

- The recently developed MRAN algorithm was also investigated in this book for nonlinear system identification. Based on run-time analysis of the MRAN algorithm in real-time identification, a new algorithm called extended MRAN (EMRAN) has been proposed. EMRAN algorithm combines a "winner neuron" strategy to the existing MRAN method, and only the parameters connecting the "winner neuron" are adapted in the on-line approximation. In this way, the computation load of the MRAN algorithm is greatly reduced and simulation studies based on benchmark problems demonstrate that the EMRAN algorithm can improve the learning speed greatly with error close to that of the MRAN.

Conclusions and Future Work 143

- In this book, MRAN is used for the first time as a fault tolerant controller for controlling a linearized F8 aircraft model.

Simulation results indicate that the MRAN controller can implement a more compact network structure with improved tracking accuracy, so that it is very suitable for real implementation of on-line control. However, since a strict mathematical proof is lacking in this approach, the stability can not be guaranteed when a pruning strategy is employed and therefore needs further work.

2. Recommendations for Further Research

Possible areas of future work that emerge from this book include:

- Design more robust control strategies and construct more efficient tuning rules for adjusting the parameters of the system.

    1. In the identification based control strategy, although number of algorithms have been developed for stable and accurate identification of the nonlinear system, when they are applied to the control strategy, there is no rigorous theory to prove the stability of the overall system and generally the convergence of the network parameters can not be guaranteed. It is imperative that the new structures for the control system be designed and the theory on the corresponding stability and robustness be developed.

    2. In the direct control strategy, a simple control structure is utilized, in which a RBFN is used to aid a proportional controller. However, it is noted that the B matrix plays an important role in this strategy, and hence it is better to use some strategies to identify this B matrix or incorporate a robust control strategy (for example, sliding mode control) to obtain good performance.

    3. In our study for adaptive control, the tuning rules are derived under the continuous frame work. In implementation, a gradient method is used to synthesis the tuning rules in the discrete domain using approximations. Alternately, the tuning rules should be derived directly in the discrete domain, *i.e.* start with a discrete model and define a discrete Lyapunov function and then derive the tuning rules.

    4. Further work on developing control strategies utilizing the feedback linearization method may be developed.

- Further research work on the MRAN algorithm and its applications should be carried out. With results obtained in this book, we are confident that the sequential MRAN learning algorithm will be useful in areas of non-linear neuro-flight control applications. Although the current pruning strategy has been demonstrated to be very powerful in nonlinear system identification, when it is applied in real-time control, it may cause instability of the overall system. Therefore future work on the MRAN algorithm could involve the following aspects.

1. Explore the MRAN algorithm in the control applications. Construct the stability theory for the MRAN's growing and pruning strategy, as well as its EKF tuning rule.

2. MRAN algorithm requires a set of threshold values which govern the dynamically changing structure of the neuro-controller network and the network's parameters adjustment. Prior to simulation, these threshold values are being assigned on a trial and error basis, and hence an optimal combination may not be attained. To be able to fully exploit the advantages of the sequential MRAN learning algorithm in flight control application, a threshold selection strategy is to be developed.

In conclusion, this book has addressed the problem of adaptive nonlinear system control by using a fully tuned RBFN. In this book, a direct adaptive control strategy is proposed and the parameter tuning rule for updating all the parameters of the RBFN controller is derived based on the Lyapunov stability theory, which guarantees the stability of the overall system. In addition, the book also contributes to improve the performance of the existing identification scheme and investigates the performance of some indirect adaptive control strategies. Substantial simulation results were presented which strongly support the proposed algorithms in the application area of aircraft flight control.

Bibliography

1. M. Agarwal, "A systematic classification of neural-network-based control ", IEEE Control Systems Magazine, Vol.17, No.2, April, 1997, pp.75-93.

2. F. Ahmed-Zaid, P.A. Ioannou, M.M. Polycarpou and H.M. Youssef, "Idenfitication and control of aircraft dynamics using radial basis function networks ", IEEE Aerospace and Electronics Systems Magazine, Vol.11, No.7, 1996, pp.5-10.

3. E. Alpaydin, "GAL: networks that grow when they learn and shrink when they forget ", International Journal of Pattern Recognition and Artificial Intelligence, Vol.8, No.1, 1994, pp391-414.

4. K.J. Astrom and B. Wittenmark, *Adaptive Control*, Addison-Wesley, 1995.

5. W.L. Baker and J.A. Farrel, "Learning augmented flight control for high performance aircraft ", Proceedings of the AIAA on Guidance, Navigation and Control Conf., pp.347-358, New Orleans, LA, 1991.

6. G.J. Balas, W.L. Garrard, and J. Reiner, "Robust dynamic inversion control laws for aircraft control ", Proceedings of the AIAA on Guidance, Navigation and Control Conf., AIAA paper 92-4329, 1992, pp.192-205.

7. L. Behera, M. Gopal and S. Chaudhury, "Inversion of RBF networks and applications to adaptive control of nonlinear systems ", IEE Proc. on Control Theory and Applications, Vol.142, No.6, Nov.1995, pp.617-624.

8. M. Birgmeier, "A fully Kalman-trained radial basis function network for nonlinear speech modeling", Proceedings of IEEE International Conference on Neural Networks, Vol.1, pp.259-264, Vienna. Austria, 1995,

9. G. Bloch, F. Sirou, V. Eustache and P. Fatrez, "Neural intelligent control for a steel plant ", IEEE Trans. on Neural Networks, Vol.8, No.4, 1997, pp.910-918.

10 J.S. Brinker and K. Wise, "Flight testing of a reconfigurable flight control law on the X-36 tailless fighter aircraft ", Proc. Guidance, Navigation, and Control Conference (submitted), Aug.2000

11 R.W. Brockett, "Feedback invariant for nonlinear systems ", Proceedings of the IFAC Congress, Helsinki, 1978.

12 D.S. Broomhead and D. Lowe, "Multivariable functional interpolation and adaptive networks ", Complex System, Vol.2, 1988, pp.321-355.

13 D.J. Bugajski, D.F. Enns, and M.R. Eigersma, "A dynamic inversion based control law with application to the high angle of attack research vehicle ", Proceedings of the AIAA on Guidance, Navigation and Control Conf., AIAA paper 90-3407, pp.826-839, 1990.

14 G.H. Burgin and S.S. Schnetzler, "Artificial neural network in flight control and flight management systems", IEEE NAECON, 1990, pp.567-573.

15 S.K. Byoung, A.J. Calise, "Nonlinear flight control using neural networks ", AIAA Journal of Guidance, Control, and Dynamics, Vol.20, No.1, 1997, pp.26-33.

16 A.J. Calise, "Neural networks in nonlinear aircraft flight control ", IEEE Aerospace and Electronics Systems Magazine, Vol.11, No.7, 1996, pp.5-10.

17 A.J. Calise and R.T. Rysdyk, "Nonlinear adaptive flight control using neural networks ", IEEE Control Systems Magazine, Vol.18, No.6, Dec.1998, pp.14-25.

18 A.J. Calise, S. Lee and M. Sharma, "Development of a reconfigurable flight control law for the X-36 tailless fighter aircraft ", Submission to AIAA Journal of Guidance, Control and Dynamics, Sep. 2000.

19 S. Chen and S.A. Billings, "Neural networks for system identification ", Int. J. Control, Vol. 56, 1992, pp.319-346.

20 C.L. Chen, W.C. Chen and F.Y. Chang, "Hybrid learning algorithm for gaussian potential function network ", IEE Proceedings-D, Control Theory and Applications, Vol.140, No.6, 1993, pp.442-448.

21 S. Chen, S.A. Billings, and P.M. Grant, "Recursive hybrid algorithm for nonlinear system identification using radial basis function networks ", Int.J.Control, Vol.52, 1992, pp.1051-1070.

22 S. Chen, C.F.N. Cowan, and P.M. Grant, "Orthogonal least squares learning algorithm for radial basis function networks ", IEEE Trans. on Neural Networks, Vol.2, No.2, pp.302-309, 1991.

23 C.Y. Chiang and H.M. Youssef, "Neural network approach to aerodynamic coefficients estimation and aircraft failure isolation design", American Inst. Aeronautics and Astronautics, AIAA Paper 94-3599, 1994.

24 D. Claude, *Everything you always wanted to know about linearization*, Algebraic and geometric methods in nonlinear control theory, Riedel, Dordrecht, 1986.

25 C.F.N. Cowan, S. Chen, S.A. Billings and P.M. Grant, "Practical identification of NARMAX models using radial basis functions ", Int.J.Control, Vol.52, 1990, pp.1327-1350.

26 G. Feng, "Improved tracking control for robots using neural networks ", Proceedings of American Control Conference, pp.69-73, San Francisco, 1995.

27 H. Gomi and M. Kawato, "Neural network control for a closed-loop system using feedback-error-learning ", Neural Networks, Vol.6, No.7, 1993, pp.933-946.

28 D. Gorinevsky, A. Kapitanovsky and A. Goldenberg, "Radial basis function network architecture for nonholonomic motion planning and control of free-flying manipulators ", IEEE Trans. on Robotics and Automation, Vol.12, No.3, June 1996, pp.491-496.

29 C.M. Ha, "Reconfigurable aircraft flight control system via robust direct adaptive control", AIAA Paper 90-2626, Sep. 1990.

30 M.T. Hagan, "Neural networks for control ", Proceedings of the American Control Conference, pp.1642-1656, San Diego, California, June, 1999.

31 E.T. Hancock and F. Fallside, *Stable Control of Nonlinear Systems Using Neural Networks*, Ph.D. Dissertation, Cambridge Univ. Eng. Dept., Cambridge, U.K., 1992.

32 W.B. Herbst, "Future fighter technologies ", Journal of Aircraft, Vol.17, No.8, 1990, pp.561-566.

33 I. Horowitz, P.B. Arnold and C.H. Houpis, "YF-16 CCV flight control system reconfiguration design using quantitative feedback theory", Proceedings of the National Aerospace and Electronics Conference, Institute of Electrical and Electronics Engineers (IEEE), New York, 1985.

34 C.Y. Huang and R.F. Stengel, "Restructurable control using proportional-intergral implicit model following", AIAA Journal of Guidance, Control and Dynamics, Vol.13, No.2, 1990, pp.303-309.

35 C. Huang, J. Tylock, S. Engel, J.Whitson and J. Eilbert, "Failure-accommodating neural network flight control", AIAA Paper 92-4394, 1992.

36 C.Y. Huang, J. Tylock, S. Engel and J. Whitson, "Comparison of neural-network-based, fuzzy-logic-based, and numerical non-linear inverse flight controls", AIAA Paper94-3645, 1994.

37 K.J. Hunt, D. Sbarbaro, R. Zbikowski and P.J. Gawthrop, "Neural networks for control system – a survey", Automatica, Vol.28, 1992, pp.1083-1112.

38 L.R. Hunt, and R. Su, "Control of non-linear time-varying systems ", Proceedings of the IEEE Conference on Decision and Control, U.S.A, 1981.

39 P.A. Ioannou, A. Datta, "Robust adaptive control: a unified approach ", Proceedings of IEEE, Vol.79, No.12, 1991, pp.1736-1768.

40 P.A. Ioannou and J. Sun, *Robust Adaptive Control*, Prentice-Hall, Englewood Cliffs, 1995.

41 A. Isidori, A.J. Krener, C. Gori-Giorgi, S. Monaco, "Nonlinear decoupling via feedback: a differential geometric approach ", IEEE Trans. on Automatic Control, Vol.26, 1981, pp.331-345.

42 T.F. Junge and H. Unbehauen, "Off-line ientification of nonlinear time-variant systems using structurally adaptive radial basis function networks ", American Control Conference, pp.943-948, U.S.A, 1996.

43 T.F. Junge and H. Unbehauen, "On-line identification of nonlinear time-variant systems using structurally adaptive RBF networks ", American Control Conference, pp.1037-1041, Albuquerque, New Mexico, 1997.

44 V. Kadirkamanathan and M. Niranjan, "A function estimation approach to sequential learning with neural networks ", Neural Computation, Vol.5, 1993, pp.954-975.

45 I. Kanellakopoulos, P.V. Kokotovic and A.S. Morse, "Systematic design of adaptive controllers for feedback linearizable systems ", IEEE Trans. on Automatic Control, Vol. 36, No. 11, 1991, pp.1241-1253.

46 T.H. Kerr, "Critique of some neural network architectures and claims for control and estimation ", IEEE Trans. on Aerospace and Electronic Sustems, Vol.34, No.2, 1998, pp.406-419.

47 M. Krstic and P.V. Kokotovic, "Adaptive nonlinear design with controller-identifier separation and swapping ", IEEE Trans. on Automatic Control, Vol.40, No.3, 1995, pp.426-440.

48 S. Lee and R.M. Kil, "A gaussian potential function network with hierarchically self-organizing learning ", Neural Networks, Vol.4, 1991, pp.207-224.

49 T. Lepikult, "Computation Methods in Linear Algebra", "http://www.cs.ut.ee/ toomas_l/linalg/lin2/ nodel.html", 2001.

50 F.L. Lewis, "Nonlinear network structures for feedback control ", Asian Journal of Control, Vol.1, No.4, 1999, pp.205-228.

51 F.L. Lewis and T.Parisini, "New developments in neurocontrol", Proceedings of the IEEE International Conference on Control Applications, pp.86-91, Trieste. Italy, 1998.

52 F.L. Lewis, A. Yesildirek, and K. Liu, "Multilayer neural net robot controller with guaranteed tracking performance ", IEEE Trans. on Neural Networks, Vol.7, No.2, 1996, pp.388-399.

53 Y. Li, N. Sundararajan and P. Saratchandran, "Nonlinear system identification using Lyapunov based fully tuned dynamic RBF networks", Neural Processing Letters, Vol. 12, No. 3, Dec 2000, pp 291-303.

54 Y. Li, N. Sundararajan and P. Saratchandran, "Analysis of minimal radial basis function network algorithm for real-time identification of nonlinear dynamic systems", IEE Proceedings on Control Theory and Applications, Vol. 147, No. 4, July 2000, pp 476-484.

55 Y. Li, N. Sundararajan and P. Saratchandran, "Stable neuro-flight-controller using fully tuned radial basis function neural network", to appear in AIAA Journal of Guidance, Control and Dynamics.

56 Y. Li, N. Sundararajan and P. Saratchandran, "Nonlinear neuro-flight-controller design using feedback-error-learning strategy", Automatica, Vol.37, NO. 8, Aug 2001, pp 1293-1301.

57 Y. Li, N. Sundararajan and P. Saratchandran, "Neuro-flight controllers for aircraft using minimal resource allocating networks (MRAN)", Neural Computing and Applications, Vol. 10, No. 2, 2001, pp 172-183.

58 G.P. Liu, V. Kadirkamanathan and S.A. Billings, "Stable sequential identification of continuous nonlinear dynamical systems by growing radial basis function networks ", Int.J.Control, Vol.65, No.1, 1996, pp.53-69.

59 Y. Lu, N. Sundararajan and P. Saratchandran, "A sequential learning scheme for function approximation and using manimal radial basis function neural networks ", Neural Computation, Vol.9, 1997, pp.1-18.

60 Y. Lu, N. Sundararajan and P. Saratchandran, "Identification of time-varying nonlinear system using minimal radial basis function neural network ", IEE Proceedings on Control Theory and Applications, Vol.144, No.1, 1997, pp.1-18.

61 Y. Lu, N. Sundararajan and P. Saratchandran, "Performance evolution of a sequential minimal RBF neural network learning algorithm ", IEEE Trans. on Neural Networks, Vol.9, No.2, 1998, pp.308-318.

62 P.S. Maybeck and D.L. Pagoda, "Multiple model adaptive controller for the STOL/F-15 with sensor actuator failures", Proceedings of the 28th Conference on Decision and Control, Tampa, FL, pp.1566-1572, Dec. 1989.

63 P.S. Maybeck and R.D. Stevens, "Reconfigurable flight control via multiple model adaptive control methods", Proceedings of the 29th Conference on Decision and Control, Honolulu, HI, pp.3351-3356, 1990.

64 M.A. Mayosky and I.E. Cancelo, "Direct adaptive control of wind energy conversion systems using Gaussian networks ", IEEE Trans. on Neural Networks, Vol.10, No.4, 1999, pp.898-906.

65 D. McGrane, R. Smith and M. Mears, "A study of neural networks for flight control ", Proceedings of the American Control Conference, pp.2511-2515, Baltimore. Maryland, 1994.

66 W.T. Miller, R.S. Sutton, and P.J. Werbos, *Neural Networks for Control*, MIT Press, Cambridge, MA, 1990.

67 A.A. Mohammad, Y.L. Abdel-Magid, "On-line identification of synchronous machines using radial basis function neural networks ", IEEE Trans. on Power Systems, Vol.12, No.4, 1997, pp.1500-1506.

68 J. Moody and C. Darken, "Fast learning in networks of locally-tuned processing units ", Neural Computation, Vol.1, 1989, pp.281-294.

69 A.S. Morse, "Global stability of parameter-adaptive control systems ", IEEE Trans. on Automatic Control, Vol.25, No.3, 1980, pp.433-439.

70 W. Morse and K. Ossman, "Flight control reconfiguration using model reference adaptive control", Proceedings of the 1988 American Control Conference, pp.2219-2224, Jun. 1988.

71 B.G. Morton, M.R. Elgersma, C.A. Harvey and G. Hines, "Nonlinear flying quality parameters based on dynamic inversion ", AFWAL-TR-87-3079, 1987.

72 M.T. Musavi, W. Ahmed, K. Chan, K. Faris, and D. Hummels, "On training of radial basis function classifiers ", Neural Networks, Vol.5, 1992, pp.595-603.

73 E.P. Nahas and M.A. Henso and D.E. Seborg, "Nonlinear internal model control strategy for neural models ", Computers and Chemical Engineering, Vol.16, 1992, pp. 1039-1057.

74 M.R. Napolitana, C.I. Chen, S. Naylo, "Aircraft failure detection and identification using neural networks ", Journal of Guidnce, Control and Dynamics, Vol.16, No.6, 1993, pp.999-1009.

75 M.R. Napolitano and M. Kincheloe, "On-line learning neural-network controllers for autopilot systems ", AIAA Journal of Guidance, Control and Dynamics, Vol.33, No.6, 1995, pp.1008-1015.

76 M.R. Napolitana, S. Naylor, N. Charles, and V. Casdorph, "On-line learning nonlinear indirect neurocontrollers for restructurable control systems ", AIAA Journal of Guidance, Control and Dynamics, Vol.18, No.1, Jan, 1995, pp.170-176.

77 M.R. Napolitano, S. Naylor, C. Neppach, V. Casdorph, S. Naylor, "Sensor failure detection, identification and accommodataion using on-line learning neural network architectures ", AIAA-94-35118-cp, 1994, pp.489-499.

78 M.R. Napolitano and R.L. Swaim, "Redesign of a feedback structure following a battle damage and/or a failure on a control surface by eigenstructure assignment", AIAA Paper 91-2626, Aug. 1991.

79 K.S. Narendra, "Neural networks for control: theory and practice ", Proceedings of the IEEE, Vol.84, No.10, 1996, pp.1385-1406.

80 K.S. Narendra and A.M. Annaswamy, *Stable Adaptive Systems*, Prentice-Hall, Englewood Cliffs, 1989.

81 K.S. Narendra and B. Balakrishnan, "Improving transient response of adaptive control systems using multiple models and swiching ", IEEE Trans. on Automatic Control, Vol.39, No.9, 1994, pp.1861-1866.

82 K.S. Narendra, Y.H. Lin and L.S. Valavani, "Stable adaptive controller design, part II: proof of stability ", IEEE Trans. on Automatic Control, Vol.25, No.3, 1980, pp.440-448.

83 K.S. Narendra and R.V. Monopoli, *Applications of adaptive control*, Academic Press, New York, 1980.

84 K.S. Narendra, S. Mukhopadhyay, "Adaptive control using neural networks and approximate models ", IEEE Trans. on Neural Networks, Vol.8, No.3, 1997, pp.475-485.

85 K.S. Narendra and K. Parthasarathy, "Identification and control of dynamical systems using neural networks ", IEEE Trans. on Neural Networks, Vol.1, No.1, 1990, pp.4-27.

86 R.C. Nelson, "Flight stability and automatic control ", McGraw Hill Book Co., New York, 1989.

87 L.T. Nguyen, E.O. Marilyn, W.P. Gilbert, S.K. Kemper, W.B. Philip and P.L. Deal, "Simulator study of stall/post stall characteristics of a fighter aircraft with relaxed longitudinal static stability ", NASA Technical Paper 1538, 1979.

88 X. Ni and S.J.R. Simons, "Nonlinear dynamic system identification using radial basis function networks ", Proceedings of the 35th Conference on Decision and Control, Kobe, Japan, 1996, pp.935-936.

89 C.B. Hong Nigel, *Neuro-Flight Controllers for Aircraft Using Minimal Radial Basis Function (RBF) Neural Networks*, Ms(Eng.) Thesis, Nanyang Technological University, Chapter 4, 1998.

90 Y. Ochi, "Application of feedback linearization method in a digital restructurable flight control system ", AIAA Journal of Guidance, Control and Dynamics, Vol.16, No.1, 1993.

91 C. Panchapakesan, D. Ralph, and M. Palaniswami, "Effects of moving the centers in an RBF network ", Proceedings of the 1998 IEEE International Joint Conference on Neural Networks, Vol.2, pp.1256-1260, 1998.

92 J. Park and I.W. Sandberg, "Universal approximation using radial-basis-function networks ", Neural Computation, Vol.3, 1991, pp.246-257.

93 J. Park and I.W. Sandberg, "Approximation and radial-basis-function networks ", Neural Computation, Vol.5, 1993, pp.305-316.

94 J.C. Platt, "A resource allocating network for function interpolation ", Neural Computation, Vol.3, 1991, pp.213-225.

95 T. Poggio and F. Girosi, "Networks for approximation and learning ", Proceedings of the IEEE, Vol.78, No.9, 1990, pp.1481-1497.

96 M.J.D. Powell, *Radial Basis Function for Multivariable Interpolation: A Review*, Clarendon Press, Oxford, 1987.

97 K.S. Rattan, "Evaluation of control mixer concept for reconfiguration of flight control system", Proceedings of the 22nd IEEE Conference on Decision and Control, Piscataway, NJ, pp.560-569, 1985.

98 S.J. Raza and J.T. Sliverthorn, "Use of pseudo-inverse for design of a reconfigurable flight control systems", AIAA Paper 85-1900, Aug. 1985.

99 R. Reed, "Pruning algorithms—a survey ", IEEE Trans. on Neural Networks, Vol.4, No.5, 1993, pp.740-747.

100 A.C. Robinson, "Total robust control — a new concept for design of flight control systems", AIAA Paper 85-1974, Aug. 1985.

101 K. Rokhaz and J.E. Steck, "Use of neural networks in control of high alpha maneuver ", AIAA Journal of Guidance, Control and Dynamics, Vol.16, No.5, 1993, pp.934-939.

102 R. Rosipal, M. Koska and I. Farkas, "Prediction of chaotic time-series with a resource-allocating RBF network ", Neural Processing Letters, Vol.7, 1998, pp.185-197.

103 G.D.E. Rumelhart and R.J. Williams, "Learning internal representations by error propagation ", Nature, Vol.323, 1986, pp.533-536.

104 D. Rumelhart and J. McClelland, *Parallel Distributed Processing*, Massachusetts Inst. of Technology Press, Cambridge, MA, 1986.

105 D. Sadhukhan, S. Feteih, "F8 neurocontroller based on dynamic inversion ", Journal of Guidance, Control and Dynamics, Vol.19, No.1, 1996, pp.150-156.

106 Tao, I.W. Sandberg, "Approximation and radial-basis-function networks ", Neural Computation, Vol.5, 1993, pp.305-316.

107 R.M. Sanner and J.J. Slotine, "Gaussian networks for direct adaptive control", IEEE Trans. on Neural Networks, Vol.3, No.6, 1992, pp.837-863.

108 S. Sastry and M. Bodson, *Adaptive Control: Stability, Convergence, and Robustness*, Prentice-Hall, Englewood Cliffs, 1989.

109 S. Sastry and A. Isidori, "Adaptive control of linearizable systems ", Electronics Research Laboratory Memo No. M87/53, University of California, Berkeley, California, 1987.

110 F. Simon, V. Kadirkamanathan, "Dynamic structure neural networks for stable adaptive control of nonlinear systems ", IEEE Trans. on Neural Networks, Vol.7, No.5, Sep,1996, pp.1151-1167.

111 P.K. Simpson, *Artificial Neural System*, Pergamon, Fiarview Park, NY, 1990.

112 S.N. Singh, W.J. Rugh, "Decoupling in a class of nonlinear systems by state variable feedback ", Journal of Dynamics, System, Measure and Control, Trans. ASME, Vol.94, 1972, pp.323-329.

113 S.N. Singh and W.R. Wells, "Direct adaptive and neural control of wingrock motion of slender delta wings ", AIAA Journal of Guidance, Control, and Dynamics, Vol.18, No.1, 1995, pp.25-30.

114 J.E. Slotine and W. Li, *Applied Nonlinear Control*, Prentice-Hall, Englewood Cliffs, 1991.

115 S.A. Snell, D.F. Enns, and Jr.W.L. Garrard, "Nonlinear inversion flight control for a supermanueverable aircraft ", AIAA Journal of Guidance, Control and Dynamics, Vol.15, No.4, 1992, pp.976-984.

116 M. Steinberg and R. DiGirolamo, "Applying neural network technology to future generation military flight control systems ", International Joint Conference on Neural Networks, Seattle, WA, 1991.

117 B.L. Stevens and F.L. Lewis, *Aircraft Control and Simulation*, John Wiley and Sons Inc., New York, 1992.

118 N. Sundararajan, P. Saratchandran and Y. Lu, *Radial Basis Function Neural Networks With Sequential Learning*, Progress in neural processing series, Vol.11, World Scientific, UK/Singapore, 1999.

119 T. Troudet, S. Grag, and W.C. Merrill, "Neural network application to aircraft control system design ", Proceedings of AIAA Guidance, Navigation and Control Conference, pp.993-1009, New Orleans, LA, 1991.

120 H. Unbehauen, *Methods and applications in adaptive control*, Springer Verlay, Berlin, 1980.

121 J.L. Wang and N. Sundararajan, "Extended nonlinear flight controller design for aircraft ", Automatica, Vol.32, No.8, 1996, pp.1187-1193.

122 K. Warwick, "The control of dynamical systems by neural networks ", IEEE/IAS International Conference on Industrial Automation and Control, pp.341-346, 1995.

123 K. Warwick, R. Craddock, "An introduction to radial basis functions for system identification– a comparison with other neural network methods ", Proceedings of the 35th conference on decision and control, pp.464-469, Kobe. Japan, 1996.

124 P.J. Werbos, *Beyond Regression: New Tools for Prediction and Analysis in Behavioral Sciences*, PhD thesis, Harvard University, Cambridge, MA, 1974.

125 B. Widrow and E. Walach, *Adaptive Inverse Control*, Prentice Hall, New Jersey, 1996.

126 D.A. White and D.A. Sofge, *Handbook of Intelligent Control: Neural, Fuzzy, and Adaptive Approaches*, Van Nostrand and Reinhold, New York, 1993.

127 W.Q. Zhang, *Control of a High Performance Fighter Aircraft by Using H_∞ Theory and Neural-Fuzzy Concepts*, M(Eng.) thesis, School of EEE, NTU, Singapore 1997.

128 Y. Zhang, P. Sen, and G.E. Hearn, "An on-line trained adaptive neural controller", IEEE Control Systems Magazine, Vol.15, No.5, 1995, pp.67-74.

Topic Index

6 Degree-of-Freedom (6-DOF), 93
Active Control Technology, 13
Adaptive Control, 1
Adaptive Critic, 7
Adaptive Inverse Control, 5
Artificial Neural Networks (ANNs), 4, 83
Auxiliary equations, 111
Auxiliary Equations, 114
Back Propagation (BP), 21
Control Configured Vehicle (CCV), 86
Conventional Controller, 106
Dead Zone, 37
Direct Adaptive Control, 7, 10
Direct Linear Feedthrough RBF (DLF-RBF), 52
Discrete Lyapunov Function, 141
Dynamic Inversion, 2, 13
Equilibrium Point, 84
Equilibrium State, 84
Error Dynamics, 32
Euler Angles, 112
Euler Equations, 112
Exact Inverse Neural Controller, 70
Extended Kalman Filter (EKF), 21, 23, 45, 47, 128
Extended MRAN (EMRAN), 26, 47, 61
Failure Detection Identification and Accommodation (FDIA), 16
Fault Tolerant Neural Flight Control, 15
Feedback-Error-Learning, 11, 85, 118
Feedback Linearization, 3, 13
Force equations, 111
Full-Fledged, 93
Full-Fledged 6-DOF Nonlinear Aircraft Model, 110
Fully Tuned RBFN, 23
Gain Scheduling, 13
Gaussian function, 20
Growing and Pruning (GAP), 26, 73
Growing and Pruning RBF Network (GAP RBFN), 38
Growing RBF Network (GRBFN), 30, 38

Grownig and Pruning (GAP), 128
Handling Qualities, 72
Herbst Maneuver, 110, 117
Hidden Neurons, 108
High α Stability-Axis Roll Maneuver, 93
Hurwitz, 33
Indirect Adaptive Control, 7
Instability, 108
Inverse Dynamics, 97
Inverse multiquadric function, 20
Inverse Neuro-Controller, 93
Kinematic Equations, 112
Least Mean Square (LMS), 21
Linear Quadratic Regulator (LQR), 86, 94
Local Nonlinearity, 107
Longitudinal Model, 71
Lyapunov Candidate Function, 33, 88
Lyapunov Stability, 99
Lyapunov Synthesis Approach, 92, 139
Maneuver, 110
Minimal Resource Allocating Network (MRAN), 24, 45, 47, 49, 128
Model Predictive Control, 7
Moment equations, 111
MRAN Neuro-Flight-Controller, 125, 128
Multi-Input Multi-Output (MIMO), 4, 38
Multilayer Feedforward Network (MFN), 7
Multilayer Perceptron, 7
Multiquadric function, 20
Navigation Equations, 112–113
Negative Semidefinite, 90
Neural Adaptive Feedback Linearization, 7
Neural Flight Controller (NFC), 76
Neural Networks, 1
Nonlinear Dynamic System, 31
Nonlinear Internal Model Control, 6
Nonlinear System Control, 8
Nonlinear System Identification, 8, 26, 69
Nruto-Flight-Controller (NFC), 93

Off-line, 4
Off-Line Training/On-Line Control, 69
On-line, 4
On-Line Learning, 85
On-Line Learning/On-Line Control, 69, 77
On-line Structural Adaptive Hybrid Learning (ONSAHL), 47, 52
Orthogonal Least Squares (OLS), 21
Pitch Rate, 72, 96
Pole-Placement, 94
Quaternion Equations, 112
Radial Basis Function Network, 1
RANEKF, 23, 48
RBFN, 1, 19, 69
RBFN Controller, 76
RBFN Identifier, 76
Reconfigurable Flight Control, 16
Resource Allocating Network (RAN), 23, 29
Restructurable Flight Control, 16
Single-Input Single-Output (SISO), 3, 38
Sliding Windows, 63
Stability-Axis Roll Maneuver, 110
Stability, 84
Stable Direct Adaptive Control, 7
Supermaneuverability, 18
Taylor Series Expansion, 86
Thin-plate-spline function, 20
Trim Condition, 105
Trim Points, 99
Tuning Rule, 34, 91
Uniform Ultimate Boundedness (UUB), 34, 89
Universal Approximation, 87, 89
Universal Approximation Proposition, 34
Upper Bound, 89
Velocity, 72, 96